BIODIVERSITY AND THE CLIMATE CRISIS

BIODIVERSITY AND THE CLIMATE CRISIS

ESSENTIAL UNDERSTANDING AND CONNECTIONS

Richard S. Feldman

Marist College

cognella®

SAN DIEGO

Bassim Hamadeh, CEO and Publisher
Tony Paese, Project Editor
Abbey Hastings, Associate Production Editor
Emely Villavicencio, Senior Graphic Designer
Michael Skinner, Senior Licensing Specialist
Natalie Piccotti, Director of Marketing
Kassie Graves, Vice President of Editorial
Jamie Giganti, Director of Academic Publishing

Printed in the United States of America.

cognella® | ACADEMIC PUBLISHING
3970 Sorrento Valley Blvd., Ste. 500, San Diego, CA 92121

CONTENTS

PREFACE

A s a professor of environmental science and policy, two areas that have become increasingly important for me to convey to my students are a) an appreciation for the diversity and beautiful details of the species with which we share the planet and b) an understanding of how the Earth's climate is rapidly changing and the implications that has for societies and other species. A few years ago, I decided to teach a course that would combine these two topics, and I sought a book that would support it. Finding no one volume written at an appropriate level for an undergraduate course, I used one book on climate change and another on biodiversity. This anthology combines the two topics and provides a format that can support a course that partially or entirely focuses upon these topics. Most of the chapters are strictly informative, while two describe controversies and advocate for action. Although there is at least one other larger volume that combines the topics of climate change and biodiversity, this text will be most appropriate for mid-level undergraduate courses.

This anthology includes chapters from several books; therefore, at times the authors refer to other chapters in the book in which their chapter originally appeared, or they refer to that book in its entirety. Hopefully these instances will not be confusing when you encounter them. The source books for each chapter are listed, and they are worthy of your attention to gain a greater breadth of understanding of the topics introduced in this volume.

INTRODUCTION

It is likely that you, as a college student, recognize to some degree that climate change is a topic of concern for many scientists, policy makers, and the public. It has been a topic in the media and taught in some classrooms. You are probably now taking a course in which you will learn more about the causes, effects, and corrective actions associated with climate change. Most discussions, articles, and media coverage address impacts upon people, communities, and their infrastructure (e.g., illness, injury, mortality and property damage associated with increasing heat, drought, and flooding). Some of the most persuasive cases for aggressively addressing climate change are made by highlighting such devastating and costly impacts upon individuals, residences, and businesses. Especially compelling is the argument that your generation and those after you will absorb an even greater impact of the changes. The present and future changes are, and will be, at a scale that is more properly called disruption—even crisis. Because of this, I have adopted the terms *climate disruption* and the *climate crisis* to better convey that it is not mere change that we are experiencing. Disruptions present greater challenges than changes; further, a crisis requires that we respond immediately and with great fortitude. Rapidly accumulating evidence, including the readings in this volume and especially more recent reports, clearly show that we are in the midst of a climate crisis. The chapters in this book concerning climate will give you a better sense of causes and effects of this intensifying problem, including upon society and other species.

Absent from most mass media reports and from general public consciousness about climate is another co-occurring global disruption: that of increasing threats to species other than humans. With most Americans living, schooling, and working in cities and suburbia, it is easy to become detached from the species that cohabit Earth with us. Yet there is great diversity of life on the planet, most of which has existed long before the advent of our own species. The biological and cultural development of our species is, in fact, closely tied to other species, and that interdependence remains even if usually hidden from your daily view and thoughts. Earlier generations had more daily contact and interaction with various wild species, which continues in some cultures and communities. An additional purpose of this book is to provide you with a general introduction to the diversity of life, also known as *biodiversity*. Chapter 1 will describe enough background about biodiversity to allow you to appreciate Chapters 7–10 that address ways that climate disruption is affecting and will affect species in a variety of ecosystems. These later chapters, therefore, bring together the two major components of this volume, helping to clarify how climate disruption impacts not just people and societies but also our fellow global "citizens."

The first part of this volume includes two chapters concerning biodiversity—one that provides an overview and the other that presents an essay on dilemmas confronting us written by one of the foremost senior American scholars of ecology and conservation George Woodwell. In addition to his standing in this profession, I included his writing because his lecture on climate change that I attended in 1986 at Binghamton University heightened my awareness of this topic at a time when it was not widely studied by ecologists and environmental scientists.

The second part includes four chapters concerning climate disruption, starting with an essay by science historian Naomi Oreskes. She leads us through the reasons we should trust the scientific evidence that Earth's climate is changing more rapidly than would occur without human influence. The next three chapters describe such evidence, including essential features of climate, how it is changing, and some documented effects.

The third part presents the synthesis of climate disruption effects on biodiversity through four chapters. They were selected to provide some variety in geography and types of species considered.

The book concludes with a thoughtful chapter by Lord Nicholas Stern, an economist who has held positions in the World Bank and the British government where he released an exhaustive and influential report in 2006 on the economic impact of climate change. This chapter provides tangible direction and encouragement for action.

There is a geographical bias toward North America in the readings with the exceptions of chapters one through three and nine and ten which are more global in their coverage. The selection of chapters five through eight was partially guided by gaining some geographic diversity across the contiguous United States and some lands across our borders.

An anthology of this modest size is necessarily limited in its scope of coverage of topics that are very broad. The selected chapters are intended to provide introductions and examples rather than comprehensive coverage. The literature cited in each chapter can be studied to gain a greater depth of knowledge of the topics discussed. Each chapter is followed by a few questions to help draw out salient aspects of the reading or to apply the knowledge gained. Ultimately, the overall goal of this volume is to inform you about two essential and threatened features of our planet—its life and its climate—and to help you use that knowledge in your personal and professional development as an environmental steward, whether or not you become an environmental professional.

PART I

DIVERSITY OF LIFE

CHAPTER 1

Introduction to Biodiversity

Richard S. Feldman

The Earth holds an incredible diversity of life forms—on the surface and in the depths of the oceans, in grasslands, deserts, marshes, and various types of forests. Collectively, this biological diversity is called *biodiversity*, usually described as the number of species in a given area. This variety of species has been appreciated and depended on by humans for all of our 300,000-year existence, and indeed, has made our survival possible. But now **we are in the midst of a crisis, known as the sixth mass extinction**: the world is losing species at rates faster than when dinosaurs went extinct. This time it is not because of an asteroid collision with Earth, and it is also not occurring over thousands of years, but rather over centuries and decades. In this chapter, you will learn about the wonderful diversity of animals, plants, and other types of organisms, and the inherent and societal values that are at risk. Addressing every threat to species and conservation strategy is beyond the scope of this chapter; however, later chapters describe present and anticipated threats to species posed by climate disruption and describe some conservation approaches.

Opening Case Study

The Monteverde Cloud Forest is the definition of lush. To first enter the forest within the Monteverde Cloud Forest Reserve is to enter a cathedral—dimly lit, thickly walled in tapestries of vegetation, the air filled with fragrances of plants and moist earth; trees draped in greenery, hanging gardens of plants almost obscuring the bark of the trees; moss and vines hanging from thick limbs, plants rooted in the moss and on bark. The stillness is punctuated by exotic birdcalls like those of the bellbird with its creaky-gate sound. Howler monkeys chant and hoot. Along the trails is a feast of color for the eyes—flowering plants of rich colors and beautiful form. At the right time of year, the patient viewer is also rewarded with a sighting of the resplendent quetzal with its iridescent plumage. Within this forest are species that exist nowhere else.

Until the 1950s, this beautiful tropical forest was little known to those outside of Costa Rica. In 1951, Quakers from Alabama arrived in Monteverde ("green mountain") after an arduous three-month journey through Mexico and Central America. As pacifists, these 44 individuals came to a country that had no military, and they bought land to farm and to produce dairy products. A few local families were already doing the same, largely isolated from the rest of the nation. The locals and the newcomers worked together and gradually built a small town. The Quakers set aside 554 hectares as a hydrological preserve to assure a water supply. In the 1960s naturalists were visiting the area and began documenting the unique flora and fauna of the cloud forest in this montane (mountain) region. In 1972 a young graduate student, George Powell, came to Monteverde to conduct

Figure 1.1 Dense and diverse vegetation in the Monteverde Cloud Forest Reserve.

Photograph by Richard Feldman.

research for his doctoral dissertation about birds. Enthralled with the diversity of the area and appalled at the deforestation occurring by those without land title, he sought to protect the cloud forest. He joined forces with Wilford Guindon, a member of the Quaker community, to create the Monteverde Cloud Forest Reserve through a number of deals and joint efforts. Over the years many conservation and education organizations helped the preserve grow as additional lands were donated and purchased.

Since the 1980s Monteverde has grown in popularity among tropical biologists, educators, and students seeking to learn and explore in this remarkable environment. It has also become a tourist destination for those willing to make the four-hour bumpy ride up the mountains. Through many research projects, much has been learned about the biological diversity and environment of this cloud forest. Detailed studies of individual species reveal intricacies of their biology, behavior, and interactions with other species. Simultaneously, the human community and the reserve are examples of sustainable management of resources and cooperation—a model to be emulated elsewhere. Yet, despite this model and its successes, and despite the remarkable beauty and diversity, Monteverde is threatened. The threats are beyond its borders and have already claimed its most noted "citizen": the golden toad. Whether more species will be lost will depend on forces bigger than Monteverde's cooperative, diverse community and supporters and will tell us how well society can manage to reduce the multiple threats to biodiversity.

The Monteverde Cloud Forest Reserve is an example of how an area with significant biodiversity value was recognized, how people and organizations came together to protect it, and how its protection is now also dependent upon forces beyond its borders in the international effort to slow climate disruption.

Part I. Describing Biodiversity

How many species are there? Which types of organisms have the most species? Where are the numbers of species greatest? These questions help us to focus on some of the important characteristics of biodiversity. Some amount of biodiversity exists everywhere; some areas are known for their especially high number of species and some contain species that are found nowhere else. In this chapter, we will examine some important patterns of biodiversity around the world and in the United States.

Number of Species

Exploring and organizing the great diversity of life is analogous to exploring and organizing the diversity of consumer products. If you were a person from the 18th century who time travelled to present time and landed in a big box retail store, you would be overwhelmed by the abundance and variety of items before you. It would appear to be a jumble of things. However, a store manager would show you around and point out the clear organization of the floor plan with signs designating major sections like toys, automotive, garden, electronics, appliances, clothing, pharmacy, food, cosmetics, housewares, cleaning, and perhaps a few others. Within each of these sections, aisles would be labelled with subsections like women's, men's, and children's within the clothing section. Since you arrived in spring and your young daughter needs a short-sleeved shirt, you would seek the girls' section; within girls would be pants, shirts, and underwear; within shirts would be short-sleeved and long-sleeved; within short-sleeved would be shelves or racks with different sizes; within her size would be various styles, colors, fabrics, and patterns to choose from. The very one chosen is a member of that "species" of shirt, defined by its collection of features.

The equivalent to store managers in organizing life are *taxonomists*. They help to name and organize the great variety of life. The father of the modern system for doing this was the Swedish biologist Carolus Linnaeus (1707–1778), who created what is known as the Linnaean System. An essential feature of this system, *classification,* organizes the variety of species into larger groups (think of going in reverse from the short-sleeved shirt, with it being part of ever larger groups of clothing). Another essential feature is that every species is named with a *binomial* of genus and species—e.g., *Homo sapiens*. This system is summarized in Figure 1.2.

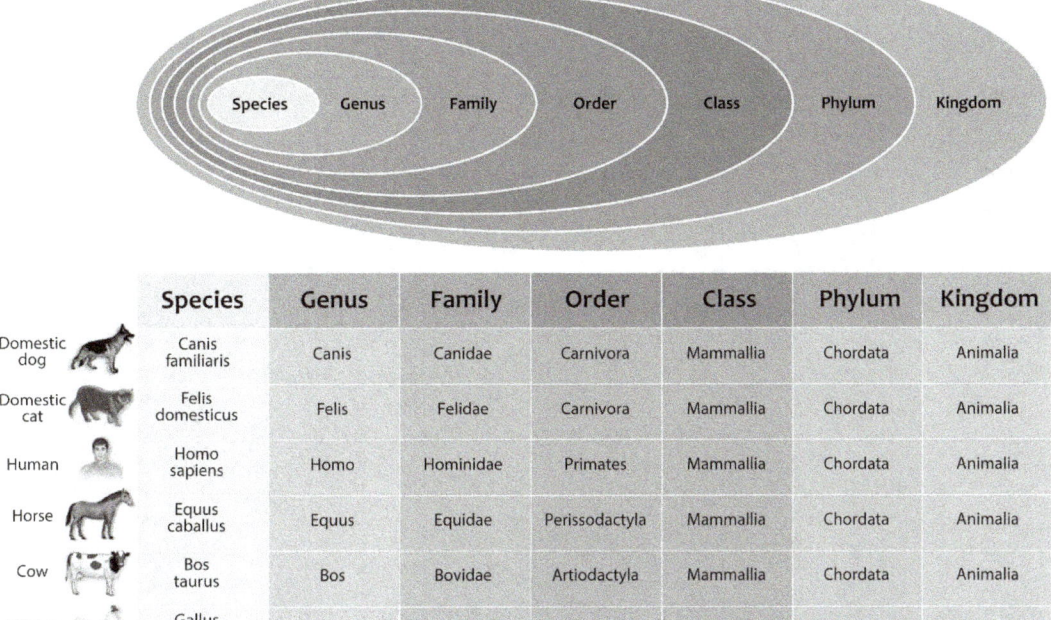

	Species	Genus	Family	Order	Class	Phylum	Kingdom
Domestic dog	Canis familiaris	Canis	Canidae	Carnivora	Mammallia	Chordata	Animalia
Domestic cat	Felis domesticus	Felis	Felidae	Carnivora	Mammallia	Chordata	Animalia
Human	Homo sapiens	Homo	Hominidae	Primates	Mammallia	Chordata	Animalia
Horse	Equus caballus	Equus	Equidae	Perissodactyla	Mammallia	Chordata	Animalia
Cow	Bos taurus	Bos	Bovidae	Artiodactyla	Mammallia	Chordata	Animalia
Chicken	Gallus gallus	Gallus	Phasanidae	Galliformes	Aves	Chordata	Animalia

Figure 1.2   Linnaean System.

There are differences in terminology used to describe the variety of species. Usually *biodiversity* is the number of different species in a place; ecologists also refer to this as *species richness*. The term *species diversity* is a combination of species richness plus *evenness*, which describes the relative abundance of different species. For example, two forests may both have 20 species of trees and therefore have the same species richness. However, if one forest has nearly equal numbers of individuals of those species it has high evenness and, therefore, higher species diversity than the other forest with unequal numbers of individuals of the 20 species. In this chapter, biodiversity will be equated with species richness.

Concept: Species are distributed unevenly across major groupings of organisms called taxa.

Different major groups of species (*taxa*) have greatly different numbers of species within them. When summarized for the world, we can see some striking numbers and patterns; Table 1.1 shows which taxa have especially high richness. Perhaps the most surprising result is that invertebrates—the animals without backbones, the non-vertebrates—are so much more diverse than any other taxon. Further, within this very diverse group, insects have by far the most species

Table 1.1   Known numbers of species in major taxa[1]

DOMAIN	KINGDOM	PHYLUM	SUBPHYLUM	CLASS	NO. OF DESCRIBED SPECIES
Archaea					377[2]
Bacteria					9,980[2]
Eukarya					
	Animalia				1,296,192
		Annelida (segmented worms)			14,399
		Arthropoda			1,082,297
				Arachnida (spiders, etc)	72,925
				Diploda (millipedes)	15,234
				Insecta	927,346
				Malacostraca (mostly crustaceans)	35,856
		Chordata			
			Vertebrata		
				Actinopterygii (ray-finned fish)	32,513
				Amphibia	6,439
				Aves (birds)	10,356
				Elasmobranchii (sharks & rays)	1,226
				Mammalia	5,852
				Reptilia	10,233
			Tunicata		
				Ascidiacea (sea squirts)	2,925
		Bryozoa			5,434
		Cnidaria (anemones, coral, jellyfish)			11,151
		Echinodermata (sea stars, urchins)			6,828
		Mollusca (bivalves, snails, squid)			65,442
		Nematoda (roundworms)			3,455
		Platyhelminthes (flatworms)			18,616
		Porifera (sponges)			9,092
		Other invertebrates			160,544
	Chromista (algae, some fungi, Foraminifera)				23,487
	Fungi				135,110
	Plantae				366,474
	Protozoa (flagellates, slime molds)				2,720
				Total	1,296,569
					>1,800,000[3]

[1] Source: http://www.catalogueoflife.org/annual-checklist/2019, accessed 7/4/19.

[2] These domains are likely to yield vastly greater numbers of species as they are more thoroughly studied.

[3] Although the figures in this column, without duplication, total to 1,296,569, the Catalogue of Life lists total species as >1.8 M.

with 927,346 identified. Notice, too, that this value represents about 50% of the total number of species that have been identified, at more than 1.8 million. There are conservative estimates that insects may number 5.5 million species (Stork 2018), and that total species may number 12 million. At a finer level of study, the taxonomic order Coleoptera—beetles—represent more than one third of all known insects (www.catalogueoflife.org). Indeed, when esteemed British biologist J. B. S. Haldane was asked what he had learned about God after a long life of studying nature, he replied that he must have had "an inordinate fondness for beetles" (Hutchinson 1959 in Laverty *et al.* 2008).

Concept: Species are distributed unevenly across geography.

In addition to the great variation of biodiversity among taxa, biodiversity also varies greatly according to location. You may have already guessed or known that different parts of the world have more species than others. Tropical areas tend to have more species than areas farther away from the equator. In fact, there is *a latitudinal diversity gradient* shown for many taxa, such as for the number of breeding bird species in the northern hemisphere (Table 1.2). This pattern operates in the southern hemisphere too, with species richness increasing toward the equator, and for a multitude of classes of organisms.

Table 1.2   Latitudinal diversity gradient among breeding birds

	LATITUDE	MEDIAN NO. OF BREEDING SPECIES
Greenland	75° N	56
Labrador	55° N	81
Newfoundland	49° N	118
New York State	43° N	195
Guatemala	15° N	469
Ecuador	2° S	1,642

Biodiversity can be described for areas as small as individual trees and their associated insects, an ecosystem like a pond, all the way up to the entire Earth. Studies of biodiversity will usually focus on a geographic area like a river valley, county, province, or country. They will also focus on a taxon like fish, birds, butterflies, flowering plants, or moss. At the end of the study, researchers are able to conclude that there are, for example, 88 species of fish in Lake Superior and its tributaries. Such studies allow them—environmental managers, students, and the general public—to know what species exist in a place and understand that biodiversity varies in their region, country, and the world.

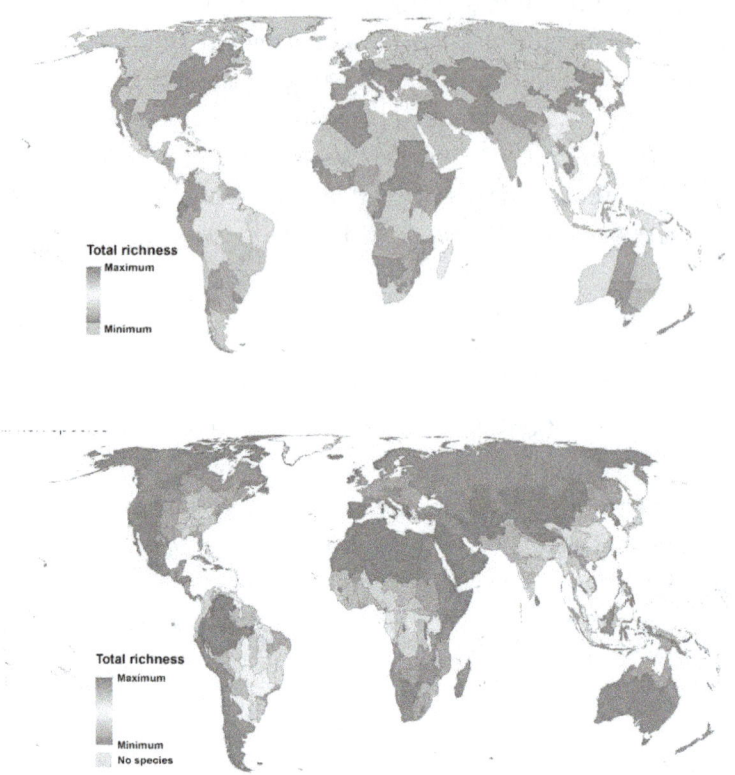

Figure 1.3 a) Flowering species diversity by region and b) Freshwater fish diversity by drainage basin.

The maps in Figure 1.3 show the variation in species diversity for a large subset of plants—those that flower—and freshwater fish. The variation in colors among nations, regions, and drainage basins indicate that the species richness greatly varies geographically. Some regions tend to have higher species diversity of flowering plants and fish than others—for example, tropical and subtropical South America, Central Africa, and parts of Southeast Asia. This is known for other taxa too like birds and mammals. Within nations there are areas of especially high diversity. For example, in the United States mammal diversity is highest in the Southwest, while tree diversity is highest in the Florida panhandle (Figure 1.4). Determining which areas have high diversity is essential for developing strategies to protect biodiversity. It is part of the important concept of *biodiversity hot spots*—areas of high diversity that are at high risk of disappearing.

You already know that many species live in the wild only in certain places. This is obvious for Australia's marsupial mammals, like kangaroos, and many of Africa's mammals of the savanna,

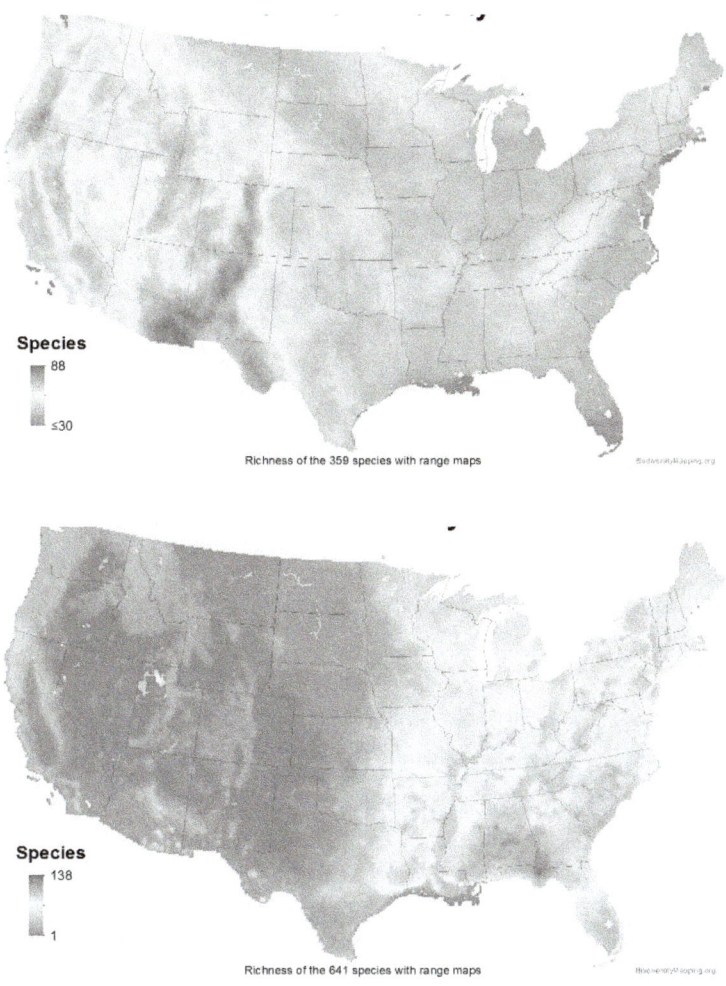

Species
88
≤30

Richness of the 359 species with range maps

Species
138
1

Richness of the 641 species with range maps

Figure 1.4   a) Mammal diversity and b) Tree diversity in the contiguous U.S.

like elephants, zebras, and gazelles. Species that are limited to individual countries, states, provinces, and even smaller areas are called *endemic species* or *endemics*. Often they are named for the place where they are found—like the Barton Springs salamander in Texas. Areas with a high number of endemics have many unique species found nowhere else in the wild. In the contiguous United States, the Southeast has especially high numbers of endemic mammal species (Figure 1.5a). Mammal endemism is especially high in Australia and Indonesia (Figure 1.5b). These areas, like those with high biodiversity, also are important in determining which areas should receive special attention for protection.

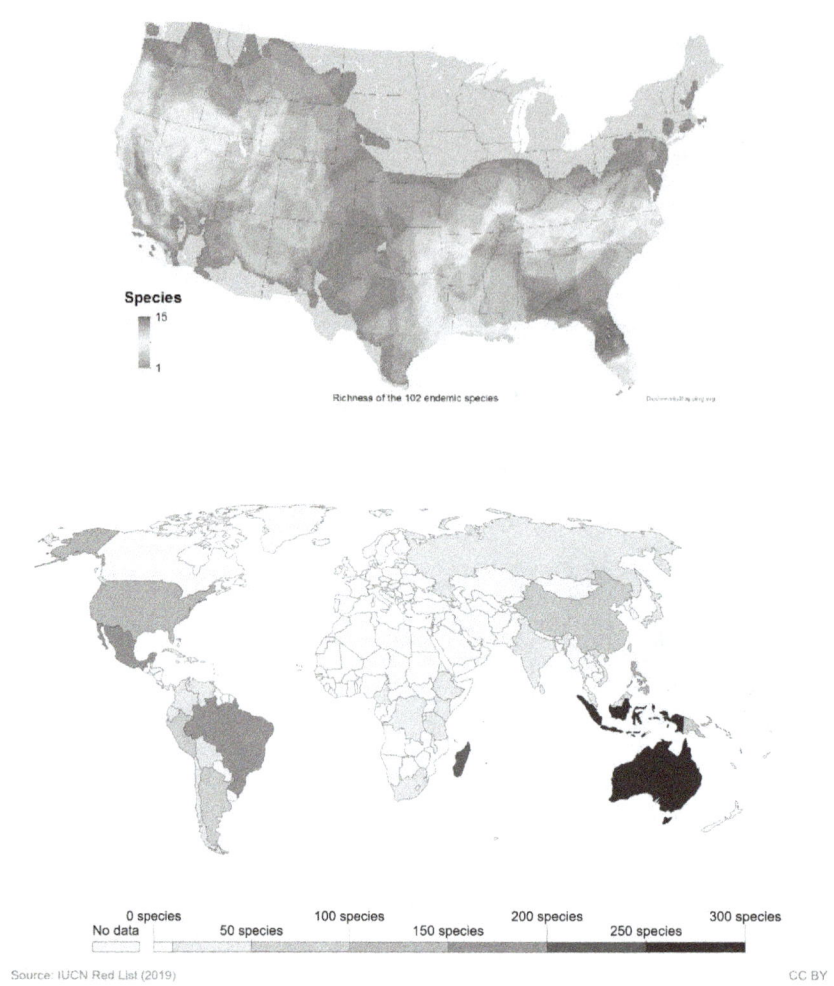

Figure 1.5   Number of mammal species unique to each a) contiguous U.S. state and b) nation.

High Diversity Regions

Concept: High terrestrial biodiversity is strongly related to abundant moisture, sunlight, and year-round warmth.

The latitudinal diversity gradient concept provided an introduction to where species richness is highest. Also, from Figure 3 you can see that parts of Latin America, Africa, and Asia rank highly in diversity. These regions have tropical moist forests, which include cloud forests like Monteverde described in the opening of this chapter and a few types of rain forests. In general, the greatest terrestrial diversity occurs in tropical moist forests. They are limited to about 7% of the Earth's land surface in the Americas, Africa, Asia, and Australia (Figure 1.6) with estimates suggesting 50% of all species live there. The peak of world plant diversity is the combined flora of Colombia, Ecuador, and Peru, and 30% of the world's bird species are in the Amazon basin.

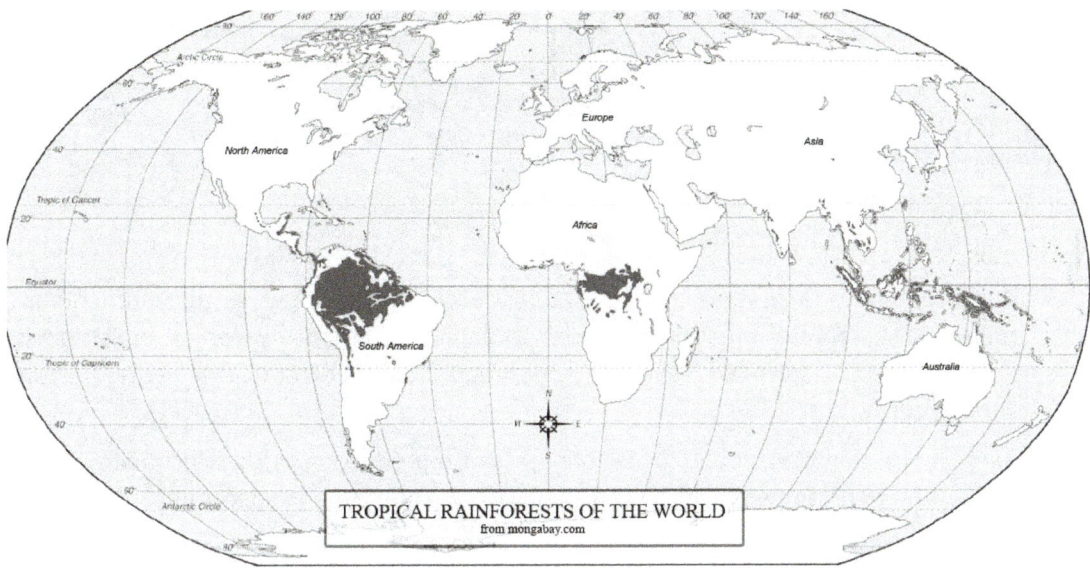

Figure 1.6   Distribution of tropical moist forests.

Comparisons of biodiversity between tropical moist forests and northern latitudes are illuminating and remarkable (Table 1.3). In the Peruvian Amazon and in Brazil's Atlantic Forest, tree diversity per hectare (two football fields) is about half of what is found for all native tree species in the United States and Canada combined! Further, the Amazon as a whole contains at least 10,000 tree species (ter Steege et al. 2019). One hectare in Panama's rain forests contains two-thirds of the beetle diversity of all of the United States and Canada. One tree in the pea family in the Peruvian Amazon was habitat for as many species of ants as exists in all of the British Isles! These are startling differences in biodiversity between the moist tropics and the temperate zone.

This remarkable diversity is supported by several factors:

- The *climate* is superb for plant growth and reproduction: year-round warmth and abundant sunlight plus high levels of rain or cloud drip (typically 100–400 in/year; cf Seattle 37", Boston 44", Mobile 67"). This supports a high diversity of plants, thereby supporting more herbivores, which in turn support more carnivores and omnivores; this tends to increase the number of species and not just the abundance of organisms.

Table 1.3   Comparison of number of species of selected taxa between Latin American tropical rainforests and North America or the British Isles

TAXON	TROPICAL RAINFORESTS	NORTH AMERICA & BRITISH ISLES
trees	300/ha in Peruvian Amazon 476/ha in Brazilian Atlantic Forest	700 native species in all of the United States and Canada
butterflies	1209/55 sq km in Peruvian Amazon	440 in entire eastern North America
beetles	18,000/ha in Panama	24,000 in all of United States and Canada
ants	43 on one tree in Peruvian Amazon	43 in all of British Isles

Sources: Gentry (1988) for trees in Peruvian Amazon; Wilson (1992) for all others.

- These conditions also allow *decomposition* to occur at high rates so that nutrients are quickly freed from dead matter for uptake by plants. Indeed, many tropical moist forests exist on infertile soils but have high nutrient availability from *rapid nutrient cycling through decay*.

- Warm, moist climates support faster growth and reproduction rates, which increase the rate of mutations, some of which are beneficial and contribute to speciation.

- Related to the preceding, as reproduction rates increase (or inversely as generation times shorten), the rate of natural selection increases, also contributing to a higher rate of speciation.

- The *lack of major regional disturbances*, like glaciation, over geologic time has allowed evolution to proceed uninterrupted for much longer periods than in many other regions of the world. Together with conditions that enhance growth, long evolutionary periods encourage the development of more species.

- The physical complexity of the forest helps to support biological complexity.

Forest stratification (multiple forest layers) is typical (Figure 1.7), starting with low ground vegetation, shrubs and ferns, sub-canopy trees, canopy trees, and finally emergent trees above the canopy. Vines and epiphytes (plants growing on the surface of other plants) create *additional surfaces and physical complexity* (Figure 1.8). These all create an abundance of habitats for animals ranging greatly in size, behavior, and ecological relations. In sum, the physical and biological diversity of the plants ends up helping to create and support the diversity of animals from tiny invertebrates to large mammals and birds.

Figure 1.7   Stratification (layering) of tropical rainforests.

Photograph by Richard Feldman.

The United States

Concept: Although the moist tropics possess the greatest biodiversity, remarkable species diversity is found in our nation as well with substantial variation in diversity across the states and regions.

The United States is notable for its diversity of certain taxa. For example, bees reach their greatest diversity at mid latitudes, not in the tropics, and the United States has nearly 4,000 species. Four freshwater invertebrate taxa reach their highest known diversity in the United States (Table 1.4).

Figure 1.8   Epiphytes and vines on one tree in Monteverde Cloud Forest, increases physical complexity and biological diversity.

Photograph by Richard Feldman.

Table 1.4   Freshwater invertebrates of especially high diversity in the United States

TAXON	NO. OF SPECIES	% OF WORLD TOTAL
Mayflies	590	30
Stoneflies	610	40
Mussels	300	29
Crayfish	322	61

Source: Stein et al. (2000).

Of the crayfish species, 96% exist only in the United States, which is one of the highest levels of national endemism for any group of organisms. Amphibian diversity is especially high in the southern Appalachians (Figure 1.9), which is attributed to its high moisture, extensive forests, and stable landscape over geologic time. The United States has the most species of salamanders of any country, with 140 species described — 40% of the world total.

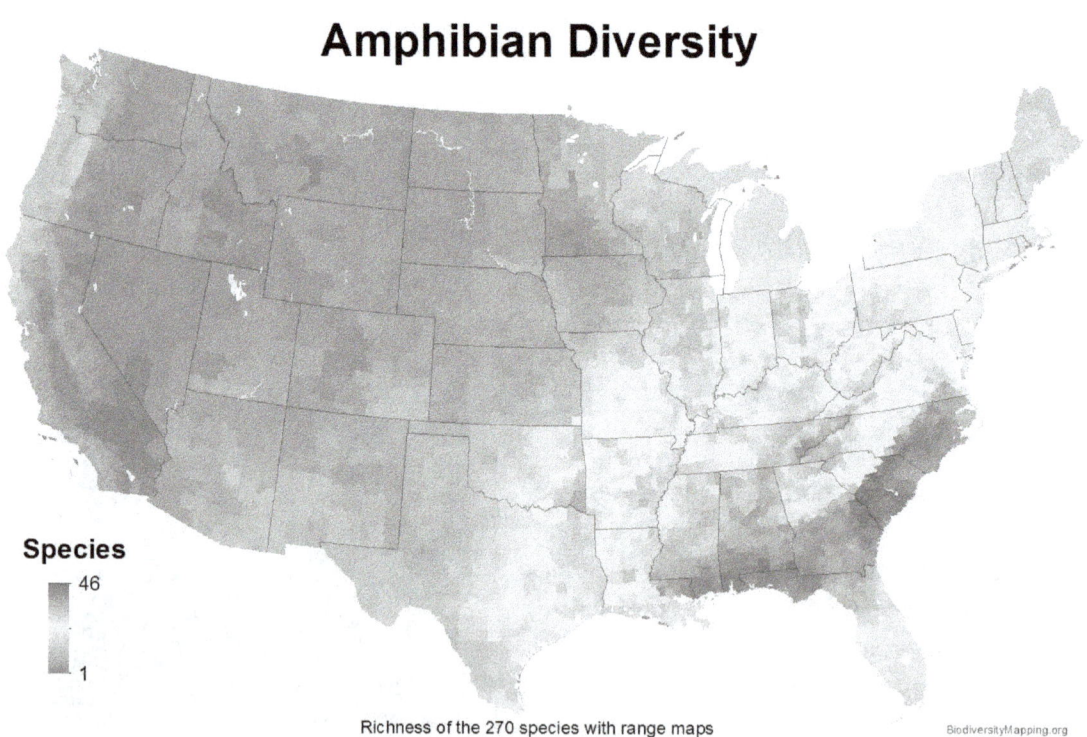

Amphibian Diversity

Species
46
1

Richness of the 270 species with range maps

BiodiversityMapping.org

Figure 1.9   Amphibian diversity in the United States reaches its peak in southern Appalachia.

Part II. Valuing Biodiversity

Now that you have an introduction to what biodiversity is, it is fair to ask why we should care about it. In modern society it is easy to not regard other species as especially important, so it is helpful to consider ways that humans benefit from both individual wild species as well as biodiversity as a whole. Among the eleven values presented below, hopefully there will be some that are especially important or persuasive for you. Ultimately, these values help to provide justification for the protection of wild species and their habitats.

Concept: Collectively and individually, species provide a vast array of benefits to human societies, whether directly for their use or indirectly for long-term human welfare as well as for the environmental stability of the planet.

Ecological

The vegetation and algae of a given region are well-adapted to the physical conditions there, producing oxygen for the atmosphere and water bodies and absorbing carbon dioxide. This absorbed carbon dioxide helps to maintain the global climate and to build carbon compounds that support food webs. In forests, the diversity of plants enables some species to thrive under the shade of others which thrive in full sunlight. As plant diversity increases so does the diversity of animals, especially herbivores and pollinators. Many of these pollinators are important for agricultural crops, as well as for wild plants, that are used by humans. The production of fruits through pollination is an ecological process that in turn supports a multitude of wild species (plus humans) with food.

Microbial biodiversity (bacteria, fungi, protozoa, algae) has important roles in nutrient cycling and decomposition. The nitrogen cycle is especially dependent on a variety of bacteria; some of these are free-living (e.g., in soil, sediments, and water) while others are symbiotic with other microbes and plants. The productivity of agricultural and non-agricultural soils is strongly dependent on bacterial and fungal populations. These are not disease microbes; they break down organic waste products and enhance the uptake of nutrients, especially via mycorrhizal fungi. In their absence, farmers must add fertilizers.

Many invertebrates are detritivores, working with decomposer microbes during decay of organic matter. They process incredible masses of debris that would otherwise accumulate and smother ecosystems, including where humans live. Many invertebrates are parasites, which help to control other populations, although unfortunately some also cause human disease and suffering. Invertebrate and vertebrate predators also help to control populations of many species, including of herbivores (like some agricultural pests) and disease vectors (like mosquitoes). Give thanks to the invertebrates!

Economic

All of our crops and livestock have their origins from wild species, many of which have been domesticated and bred over hundreds and even thousands of years. Wild relatives of our crops and livestock remain a source of genes for improvement of our domesticated foods. Many wild foods continue to be important direct sources of food, notably fish. In many rural communities hunting of animals and foraging of wild plants are primary sources of food.

Pollinators, especially insects, provide a tremendous economic benefit through the fruits and vegetables that they help to produce. Most pollinators are wild species and far more numerous than honeybees brought in to augment wild pollinators. In addition to thousands of bee species, pollinators of economic importance include flies, beetles, and butterflies.

Naturally occurring predators, parasites, diseases, herbivores, and competitors have important roles in suppressing populations of pests. Pest control affects crops, livestock, human health, and landscapes. Many species of birds, bats, and predatory insects feed upon biting insects, including some that are disease vectors (carriers), like some species of mosquitoes. Diverse and thriving soil microbe communities help to out-compete soil-borne plant pathogens, while diverse soil invertebrate communities help to suppress crop and livestock pests. Predatory vertebrates like hawks, other raptors, and snakes feed upon rodents that, when overpopulated, become significant agricultural pests and human disease vectors. Some herbivores and weed species change the plant species composition of landscapes and damage private properties. These can often be controlled through conservation and reintroduction of native competitors, predators, and herbivores. From Table 1.5 you can see 13 categories that relate to food and the dollar values associated with them. These are extraordinary economic benefits of biodiversity that are taken for granted by most people.

Biodiverse areas attract visitors—maybe even you. Have you visited places that were especially interesting or beautiful because of the diversity of life? For example, coral reefs are major destinations for snorkelers and divers. Animal enthusiasts seek out locations with a variety of vertebrates and even invertebrates. Flower lovers seek out locations known for their diversity and unusual species. Hunters and anglers often seek areas with diversity as well as abundance of preferred prey. Many of these individuals spend money related to their activities, and some of those expenses support local economies in the biodiverse areas, providing a direct economic motivation for protection of the species and their habitats. Table 1.5 lists estimates of the dollar values attributed to elements of ecotourism and other activities which often contribute substantially to the economy of developing nations.

Biodiversity plays an important role in ecosystem functions that provide services essential for human well-being. For example, protecting the Catskill Mountain watersheds that provide drinking water for New York City paid a dividend of $10 billion by avoiding the cost of a water treatment facility. The diversity of species in those protected forest ecosystems have interdependent roles that contribute to maintaining water quality and quantity in streams and reservoirs. They effectively replace very expensive water treatment facilities that would have to be built and

Table 1.5   Total estimated economic benefits of biodiversity in the United States and worldwide

ACTIVITY	UNITED STATES (× 10⁹)	WORLD (× $10⁹)
Waste disposal	62	760
Soil formation	5	25
Nitrogen fixation	8	90
Bioremediation of chemicals	22.5	121
Crop breeding (genetics)	20	115
Livestock breeding (genetics)	20	40
Biotechnology	2.5	6
Biocontrol of pests (crops)	12	100
Biocontrol of pests (forests)	5	60
Host plant resistance (crops)	8	80
Host plant resistance (forests)	0.8	11
Perennial grains (potential)	17	170
Pollination	40	200
Fishing	29	60
Hunting	12	25
Seafood	2.5	82
Other wild foods	0.5	180
Wood products	8	84
Ecotourism	18	500
Pharmaceuticals from plants	20	84
Forests' sequestering of carbon dioxide	6	135
Total	$319	$2928

Source: Pimentel et al. (1997).

paid for to maintain drinking water quality. Simultaneously, that watershed protection has also helped to protect a beautiful landscape, streams and reservoirs that draw increasing numbers of visitors who help to support the local economy through their purchases and lodging.

Medicinal

Many medicines originally came from nature, whether from organisms or minerals. Plants, animals, and microbes remain an important source of new drugs, even if their compounds can be later synthesized in labs and factories. Aspirin, digitalis, and penicillin are examples of common drugs originally derived from plants and fungi. More recent discoveries include paclitaxel from the bark of Pacific yew, found in forests of the Pacific Northwest, which is used to treat ovarian and advanced breast cancer (Figure 1.10). The Madagascar periwinkle yields vincristine and

Figure 1.10 Medicines from nature a) Madagascar periwinkle; b) Pacific yew yield cancer treatments; c) Cone snails yield painkillers.

vinblastine, which are used to treat various lymphomas, advanced breast cancer, and advanced testicular cancer. Ziconotide painkillers come from cone snails.

Typically, researchers learn of the use of a plant by local herbalists and traditional health practitioners, and after chemical analysis, they recognize that certain compounds may have additional benefits. That's what happened with the wild yam, which was originally used to treat menstrual cramps and menopausal symptoms. Its tubers contain diosgenin, a steroid-like substance that is used in the industrial synthesis of several hormones, including progesterone, which is important in controlling human fertility. Consequently, it had a key role in the development of the birth control pill during the 1950s.

Evolutionary

What music are you listening to these days? Who do you suppose those musicians listened to and were influenced by? Keep on thinking back in time about who the influences were of each preceding musician. Where does it take you? Likely back to early rock & roll musicians and then back to various other musical heritages, such as jazz, gospel, folk, bluegrass/old time, and classical. Each of those heritages had their major players. Now, imagine if one or more of them never lived or died before they produced their influential pieces? How would music have developed after them? For example, imagine if Elvis Presley and Ella Fitzgerald never existed and all the musicians who came after them never felt their influence. What would today's music sound like?

Just as the music we hear today developed out of the music from the past, it is true that the species on Earth today evolved from species that existed before them. All of today's species have predecessors; if some of these had prematurely gone extinct, certain species could not have evolved from them, and different species would be present today.

Now, think about music again: what will it sound like during your lifetime? In ten years? ... 30 years? ... 60 years? It's hard to know, but this much is sure: it will be influenced by the music that precedes it. Next, what species will be on Earth in the future? Again, hard to know, but it is completely dependent upon which species are present for the forces of evolution to work upon. If species are removed, they certainly cannot play a role in future development of species in the process of evolution.

So, each species has a role to play in the ongoing process of evolution, wherever it leads. The removal of species (extinction) forever changes the course of evolution.

Symbolic

Many species are symbolic for cultural, political, religious, and other reasons. Sports teams and schools use animal mascots—do you remember yours? State birds, flowers, and trees are designated for each state of the United States—what are they for your state? National and state emblems and flags often include species. Eagles, bears, and lions are commonly used to depict strength and courage, while doves and olive branches represent peace. The staff of Asclepius, with a snake entwined around a tree limb, is a universal sign of medical practice. There is some irony that our national symbol, the bald eagle, had been on the Endangered Species List for more than three decades until 2007. This example also serves symbolic value in how well-informed and concerted action leads to environmental success.

Aesthetic

There is a great allure to spotting and hearing animals in the wild. For birders, the sounds and sights bring joy and appreciation of nature's beauty and variety. For fish anglers it includes admiring the sleek form, colors, size, and other details of a fish just caught. Painters and photographers bring out the details of plants and animals or use them to inspire their creativity. The array of species in an area—its biodiversity—is also appealing. For example, the dazzling assortment of color and form in an alpine meadow in full bloom, of shifting form and color on a coral reef as tropical fish come and go, and the subtle hues and textures on an autumn forest landscape in the Northeast. Stroll through a forest on a moist spring morning, hearing the diversity of birds above, smell the complexity of fragrances from the soil, decaying wood, and plants, and see the assortment of flowers, leaves, and bark. That's biodiversity in its multisensory beauty. We are drawn to the attractiveness and strangeness of living forms in their uniqueness and their variety. They stimulate our curiosity, wonder, creativity, and emotions. Returning to the earlier description of the Catskill Mountains and its economic benefits, as a resident of that region I am a daily beneficiary of the aesthetic qualities of that landscape.

Spiritual

Many people believe that the role of determining what should live is reserved for a power greater than humans. For some this is called God, the Great Spirit, Mother Earth, the Creator, and many others. Determining which species continue to imbue the Earth with their qualities or pass on their genes in evolution is a role much too awesome for one species to be responsible for. For those who regard the Earth and its species as God's creation, as God's handiwork, they may question whether humans have any right to destroy what God has created. Indeed, many such people see their role as stewards of God's creation—to assist in its growth and protection. There is certainly much mandate for this in religious teachings.

Nature, and especially the variety of species and the solitude of undisturbed places, is a foundation of spiritual connection for many people. Many individuals and cultures rely upon nature as a sanctuary and a source of renewal, inspiration, and meaning for their lives. This occurs in industrialized and non-industrialized societies all around the world.

Mental Health

Have you experienced the sense of calm that often comes with entering a forest, meadow, or other natural area where life is abundant? As you spend more time in such places perhaps you have felt some of your worries and anxiety melting away. When my students complete entries for field journal assignments, describing what they had observed in the natural world, many of them also comment at the end how relaxed they felt after an hour, able to leave behind the stress of school for a while. This effect is well-recognized by a growing number of health care providers and especially by psychologists and social workers. Some colleges have organized programs to lead students into natural areas for meditation, yoga, group counseling, or just being away from the pressures of daily life. Terms like Nature Rx and Forest Bathing are used to promote the mental health benefits of spending time especially in natural areas with abundant plant life.

Recreational

Much of our aesthetic appreciation of species and of biodiversity is part of our recreation. Our leisure time is often spent seeking out certain species or enjoying biologically diverse ecosystems. These activities also have the economic component described above. Entire sectors of the recreation industry assist us in appreciating wild biodiversity (e.g., binoculars, cameras, field guides, fishing gear, hiking and camping supplies, touring companies, lodging, transportation, preserves, parks, etc.).

Every Christmas Day in the United States, avid birders visit favorite areas to record the species they see and hear. Not only is this a national day of recreation for these people, but their records become data for the over-wintering occurrence of bird species in the United States. These data then become important for education and research about bird ranges, population sizes, and how these may be changing over time.

Education and Research

Individual species and communities of species have great interest for those who wish to know more about the world around them. Their study supports all the benefits described above. The value of something tends to increase as we learn more about it, whether the value is tangible (food, profit) or not (personal meaning). Indeed, as you continue to learn more about species and their communities, you will likely appreciate their value more. The intellectual process of learning about species brings personal and professional satisfaction to many people of all ages around the world. It is part of a greater understanding of the world. For some it is also the work they do.

Inherent Value

The previous values are dictated by the perspective of one species: *Homo sapiens*. Yet species do not need to have value to humans to justify their existence. Most species have been present longer than us, so they even have seniority!

References

Gentry, A. H. 1988. Tree species richness of upper Amazonian forests. *PNAS, 85*(1), 156–159.

Laverty, M. F, E. J. Sterling, A. Chiles, & G. Cullmanl. 2008. *Biodiversity 101*. Greenwood Press, Westport, CT.

Pimentel, D., C. Wilson, C. McCullum, R. Huang, P. Dwen, J. Flack, ... B. L. Cliff. 1997. Economic and environmental benefits of biodiversity: The annual economic and environmental benefits of biodiversity in the United Sates total approximately $300 billion. *BioScience, 47*(11), 747–757.

Stein, B. A., L. S. Kutner, & J. S. Adams. 2000. *Precious Heritage: The Status of Biodiversity in the United States*. Oxford University Press, NY.

Stork, N. E. 2018. How Many Species of Insects and Other Terrestrial Arthropods Are There on Earth? *Annual Review of Entomology, 63*, 31–45.

ter Steege, H., S. Mota de Oliveira, N. C. A. Pitman, D. Sabatier, A. Antonelli, J. E. Guevara Andino, G. A. Aymard, & R. P. Salomão. 2019. Towards a dynamic list of Amazonian tree species. *Scientific Reports 9*, Article number: 3501.

Wilson, E. O. 1992. *The Diversity of Life*. Harvard University Press, Cambridge, MA.

Discussion Questions

1   What is the role of taxonomists?

2   What is the benefit of having a system that groups similar species together? Include what it shows and how it can aid in the learning and appreciating of biodiversity.

3   Which types of organisms are especially diverse?

4   Explain why the moist tropics have especially high species richness.

5   Which values of wild biodiversity do you think are especially persuasive? Explain why and to whom they would be convincing.

CHAPTER 2

The Limits of Biodiversity

George M. Woodwell

> I celebrate myself, and sing myself,
> And what I assume you shall assume,
> For every atom belonging to me as good belongs to you.
> —Walt Whitman, *Song of Myself*

The terror engendered by nuclear weapons was balanced in part during the two decades following their use in war by the widely shared anticipation of abundant cheap energy from the atom to drive industrial development globally. The focus of attention turned to governmental alliances with corporate industrial interests. It was easy at that time to slip into the view that cheap energy is a universal good, capable of use to cure virtually any political or economic ill.

I, too, was trapped in this popular view, and considered a major subsidy to nuclear businesses, the Price-Anderson Act limiting industrial and governmental liability in a serious accident, as a wise and necessary step into the new world. My experience was limited, of course, although it did at that point include two years at sea on a naval oceanographic vessel, a steel-hulled diesel "ecosystem," entirely dependent on those deep-down bunkers, tanks of diesel oil, energy. Oil made everything possible, not only mobility, but also lights, heat, freshwater, and air in confined quarters as well as compass, radio, ballast, and pumps. It was that energy, thousands of gallons of it, that kept the ship more or less stable and afloat in the North Atlantic storms that we seemed to seek out to test our mettle as we explored from above the monstrous mountains of the Mid-Atlantic Ridge, then new to science and the public. We were not a good example for we were squandering the future, heedlessly extruding our abundant wastes to float off or sink into an apparently limitless sea, one ship among thousands on the oceans with the same callous scorn for simple manners.

Despite the compelling lessons from the weapons tests, the hard news about the biotic hazards of nuclear weapons, and how their wastes accumulate and circulate widely to become hazards to all, the health of the biotic environment was, for the moment at least, set aside by science, government, and business. It was forgotten. The objective was economic growth, and the prospects seemed unlimited.

It proved to be an unfortunate oversight.

The scientific community, at least part of it, participated with enthusiasm. Democratic capitalism seemed to offer nearly unlimited possibilities in construction and engineering, and in bending the world to suit expanding human interests, aspirations, and numbers. Some biologists, however, were increasingly sensitive to the explosion of the human enterprise into a biosphere whose limits were already being tested. After all, thoughtful scholars were realizing then that our civilization spans at most a few thousand of earth's four-billion-year history. Life itself is an anomaly, and humans are a recent arrival, new and potentially short-term guests in a biosphere owned and operated by the joint efforts of tens of millions of species too numerous to count. The evolution of life over the four billion years of the planet's existence, and especially over the most recent five hundred million years, built the environment of all life as we know it. The total habitat of that life is the thin earthly surface: the atmosphere, the oceans, lakes, and streams, the shallow crustal soils, rock, and elsewhere, the ice. That small volume comprises the entire biosphere, the habitat of all the life we are privileged to know. It extends at most a few miles from the oceanic depths to the top of the highest mountains. A few spores may be carried higher in a turbulent atmosphere, and ancient microbial forms lie deep in crustal sedimentary rocks, but for all practical purposes the functional biosphere reaches its limits with Himalayan peaks and the Mariana Trench, the deepest oceanic cut in the world.

Within that realm, life is a wild scramble for survival among perhaps a hundred million separate forms, or species. Ultimately, each survives by dint of the activities of all. And all survive in a milieu of each, sorted, resorted, and formed into mutual dependencies and overlapping interests that keep them together as guilds or simple communities that define the place and environment, and maintain both—a perfect tautology in nature.

On land, the life at a location defines the site. Much of the Northern Hemisphere is forest. In the northern regions it is the boreal coniferous forest, which is circumpolar and the largest forested region in the world. In eastern North America, the eastern deciduous forest is the natural vegetation of nearly a quarter of the continent. In water, biotic guilds, dependencies as simple as krill and whales, exist but the geography dominates, and we name lakes and streams, shallows and deeps, bays and seas and currents on the basis of what is easily identified. On land and in water, these are the communities that have emerged as the operating systems of the biosphere. They are the products of the evolutionary processes that built the human habitat and maintain it now.

Many biologists and others were increasingly concerned in those postwar years that one species of the biosphere's many millions had in a few decades come to dominate, and in fact control without a plan, the future of all that life. The brutal experience with nuclear weapons showed how vulnerable all human works are—cities, nations, and millions of people—to vaporization in the anger of war. But even more significantly, the experience demonstrated that far short of the convulsion of war, the normal circulating systems of the biosphere are vulnerable to corruption by what, taken individually, may appear to be minor releases of our civilization's waste products.

Such dominance by one species would seem to put the structural and functional integrity of the biosphere at the forefront of human interests, the primary concern in building governments globally in support of a growing population and an industrially expanding, and promising civilization. The experience with nuclear weapons might have been expected to offer a clear warning to everyone just how large the human influence had become as well as how tenuous the human hold on a suitable and nurturing environment might be.

Far from it.

As the human presence expanded and economic growth seemed not just essential but also the very mark of progress, it was convenient for governments and burgeoning industries to assume that the environment would take care of itself, clean itself up after use, and require no special attention. The momentum of the political and economic forces had been such that those forces could redefine the public interest and welfare as "economic growth" at any cost, with "the environment" as a kind of luxury that could be dealt with at a later stage if necessary. Meanwhile, the incremental corruption of air and water gnawed at the panoply of human rights developed over centuries to the point where only conspicuously fatal municipal disasters such as the smog of London and Pittsburgh of the 1940s and 1950s could move governments to action in regulation.[1]

The response of many in the scientific and conservation communities was not at all surprising. More comfortable with small matters that they could control, they turned away from government and business to talk among themselves, accumulated friends who would listen, and built separate nongovernmental institutions designed both to operate independently in conservation and to pressure governments along the way. They had no difficulty in attracting interest. Human interest and imagination have ever been easily captured by the diversity of forms and functions that throng the earth.

For many conservationists, attention in the second half of the twentieth century turned to protecting vulnerable species. Experience with losses of uniquely spectacular forms had seared the human record and a systematic approach to correcting the trend seemed an obvious necessity. The extermination of the fearless and docile Steller's sea cow (*Hydrodamalis gigas*), the largest sirenian and a true giant, reported to reach as much as thirty feet in length and several tons, stands as an early scandalous tragedy, impossible to ignore. At one time individuals ranged along both shores of the North Pacific. By 1768, just twenty-seven years after the first description, the species had been hunted to extinction. Scores of other examples of extinction followed, including the infamous loss of the North American passenger pigeon (*Ectopistes migratorius*), hunted in large numbers for food until there were no more, with the last one dying in a Cincinnati zoo in 1914. In the decades following two world wars in the twentieth century the expansion of the human presence globally put many other species, including whales, fish, birds, large mammals, and various plants, and still isolated human communities in conspicuous danger.

The response of thoughtful biologists was direct: a listing of endangered species, fauna, and flora, known as the "Red List." It was a compendium of the world's most threatened species published at regular intervals by the International Union for Conservation of Nature (IUCN), an organization established in 1948 in the heady postwar years of hope and innovation. It was founded to expand conservation interests and especially to appeal to the global academic community. The Red List gave formal status to species on an international level that might be used as a basis for public action in protection.

The IUCN quickly became the icon of conservation. Its success and need for funds for urgent new projects led a number of supporters in 1961 to found the World Wildlife Fund (WWF) with an international office in Morges, outside Geneva, Switzerland. The plan was to raise and distribute funds to support IUCN initiatives in science and conservation, and develop the WWF around specific national offices ("appeals"). The US appeal, WWF-US, grew over the course of the next three decades to dominate the world organization. Not surprisingly, the national branches, ever successful in gathering funds from engaged supporters, found it attractive and even important in fund-raising to develop their own programs quite independent of the IUCN. Many innovative programs were created along the way, but the Red List concept easily emerged in the eyes of many conservation agencies as a primary statement of the mission of conservation. A listing on the Red List put the finger on species and specific sites requiring attention and ultimately preservation. And the whole program appealed to the public, which responded generously to requests for funds to save attractive animals in danger.

The approach was effective. By 1963, recognition of the need for more formal support for preserving species against rabid commercial enterprise led the IUCN to draft the Convention on International Trade in Endangered Species (CITES). The convention was designed to prevent commerce in already-known endangered species. By 1973, it had received sufficient governmental recognition that a meeting of representatives of eighty nations could be held in Washington, DC, to establish a process for ratification and implementation of the treaty. CITES entered into force in 1975 after 10 signers had ratified and become parties to the convention. By 2009, 175 nations were parties, and the convention gathered muscle as nations developed their own internal rules for protecting species, such as the United States with the passage of the Endangered Species Act (ESA) of 1973. Additional attention was by then being drawn to special international challenges such as the continued killing of whales despite the International Whaling Commission's decades of efforts to protect them. Big cats and large pelagic fish such as giant tuna and sharks were also increasingly under pressure, and had drawn special attention in conservation.

The emphasis on endangered species was attractive, understandable, practical, and effective in drawing in powerful interest and financial support. But it would often prove inadequate in the core objectives of conservation, even for the species in question. Every species has evolved into a role, large or small, in a community of species. A species threatened is a sign of a community threatened and effective conservation reaches immediately beyond individual species to the recognition of the community and its protection, a much larger challenge that immediately touches land use as well as diverse economic and political interests.

Meanwhile, in the academic world research papers and books appeared around the truism that *extinction is forever.* The Red Lists became longer. The number of species threatened became so overwhelming by the early 1970s that the challenge had been transformed into concern for all species.

A shorthand emerged. "Biodiversity" truncated the term "biotic diversity" and came to consolidate all purposes in preserving species. It leaped over the virtual impossibility of conserving each species, one at a time, and called attention to all life. Defending biodiversity in all its imaginable forms became the core objective in conservation. It was a loose and vague formulation, but economic and political powers find such vagueness often convenient, and the assumption was that governments would surely understand and cooperate. After all, the reference was to all life on earth.

The concept of biodiversity was soon bolstered by books and compendia of papers written by many of the world's best-known scholars asserting articulately the importance to all of preserving every sprig of the earth's green mantle.[2] I was an active participant, eager to advance any effort that would bring greater public recognition to the importance of preserving all the life on earth.

* * *

In 1973, Tom Lovejoy, who had completed a PhD at Yale on bird populations in the Amazon basin, joined the staff of the still-new WWF-US, where I was at that time a board member and chair of programs. There was then great interest in obtaining objective evidence that the metric of biodiversity, the number of species present, defined structure and function in landscapes, and conveyed substance to ecology and succor in various forms to human interests. The search intensified over time as scholars recognized that there are large differences in biodiversity around the globe. Tropical forests, for example, have many more species than boreal forests. "Hot spots" of high biodiversity are obviously more attractive and important as objectives in conservation than places less well endowed. Or are they?

Lovejoy had recognized that the expansion of agricultural activities in the Amazon basin might offer an opportunity to examine the changes in plant and animal species as the area of forest diminishes. In a short time he had arranged an imaginative and aggressive test. He was able to define and establish forest plots of various sizes in an area along the Rio Negro north of Manaus to be preserved while the regions around them were cleared for agriculture. Bird populations in particular were inventoried but records of other species were kept as well over years. It was a magnificent test, and ultimately showed how sensitive such forested lands are to disturbance and what an excellent index of change an inventory of birds offers. All had previously recognized that some forest dwellers such as the jaguar, for instance, range over areas of many square miles, but the sensitivity of bird populations as well as other animals and plants to disturbances had not been widely recognized; further, the demonstrations of losses of species in these experimental plots were unequivocal: preservation of diversity required large areas, thousands of hectares, left intact.[3]

Many other efforts were also undertaken by scientists over the next years to define the significance of mere diversity—the number of species present—in maintaining the structural and functional integrity of communities.[4] Ultimately, such studies would confirm the necessity for preserving large landscapes to assure the continuity of species endemic to a locale. And the species endemic to the locale represent communities whose integrity is essential to other species in the area.

The emphasis on species became a political issue stamped as important by passage in the US Congress in 1973 of the Endangered Species Act (ESA). The act was designed, as the title suggests, to protect species in danger of extinction. Congressional recognition of the importance of protecting species in danger was unquestionably a positive step. It had the unfortunate connotation that only species recognized as interesting and endangered required attention. Overlooked were the issues of land and water management critical to stabilization of essential resources, including drainage basins and coastal land and water. To politicians and the public at large it may have also reinforced the notion that conservation could be set aside as a secondary issue in political and economic affairs, to be considered only on special occasions when a species was at hazard. Then, conservation interests, at least in the United States, had to be placated. And for conservationists, there seemed no other legally powerful way to raise issues of biotic conservation before government. That was an unfortunate and inconvenient fact—a serious stumbling block to effective governance in protecting essential landscapes and water bodies. Its inconvenience and ineffectiveness led to an effort favoring far more comprehensive legislation aimed at improved environmental planning and management in the United States, at the time a world leader on such issues.

One form of this effort emerged as a special report prepared by the National Research Council of the National Academy of Sciences in 1993, twenty years after the ESA. It was a bold and appropriate, if belated, step in coupling preservation of the integrity of natural communities as opposed to species to management of common property rights to clean air, water, and a wholesome, supportive, and vital biosphere. The proposal was for the United States to establish a "National Biological Survey" similar to the US Geological Survey in the Department of the Interior. Its mission would be to identify intrinsic biotic resources and recommend management procedures for their protection. It was a much-needed innovation in government in the United States, a potential model for the world, at a time when the world was increasingly enshrouded by giant international corporate machines and the human population was rapidly closing in on a total of seven billion with the prospect of billions more.[5]

The US Congress had a different perspective, however. So many members of Congress were, one way or another, influenced by exploitive business interests that there was no serious consideration of the proposal. Further, the very success of the ESA had generated sufficient friction in the economic growth community of business politics that there was no chance that any further potential biological restrictions, even those clearly aimed at preserving common property interests such as a forested drainage basin feeding a water supply, would survive. If there were vital issues of conservation involved, they had to be seen through the lens of direct hazards to people or species under the Endangered Species Act, which many political and economic interests also opposed but had to accept. That interesting and progressive piece of legislation, cramped in context and potential as it is, has nevertheless survived, and emerged as what may be the most critical tool in the United States for managing forests and land over the years since its enactment.[6] Its great strength lies in the reality that preserving a species requires that the habitat be preserved intact. Its weakness is just that fact: a single species draws ire if it is the full reason for preventing a lucrative real estate development. The issue writes the text of the contradictions intrinsic in the destructive expansion of the human enterprise in a finite and vulnerable biosphere whose integrity of function is essential to the continuity of the human enterprise.

* * *

"Biodiversity" has, nonetheless, become part of the vernacular, used often in discussions of conservation to refer to all life in a place. Places of high biodiversity, rich in species, are of greater interest for conservation, of course. But establishing a park to protect high biodiversity in a special place is futile if the critical aspects of the regional environment erode and the park slips into progressive impoverishment. The protection of ecological integrity or "conservation" reaches far beyond a simple emphasis on protecting species or biodiversity in parks and reserves.

The regional and global environmental transitions now under way are moving the environment out from under each individual—not species as such, but each organism, each survivor of the competitive business of life, and every special genetic strain, or "ecotype," which virtually every individual represents. In such a world, saving a hot spot of biodiversity is little more than a hope, certain to be ultimately dashed if chronic changes in environment accumulate and produce the inevitable impoverishment of land and water. That set of challenges, so lucidly set before the world by the decade of nuclear weapons tests and the subsequent lessons from pesticides, places the protection of global chemistry firmly in the realm of both conservation and human welfare. Both are the business of government—perhaps its most important business.

Elizabeth Kolbert in a recent book vividly illustrates the importance of global chemistry to species preservation in describing her personal adventures in following scientists to experimental sites on Australia's Great Barrier Reef and to a carbon-dioxide-rich fumarole off the coast of

Italy. The research she depicts confirms powerfully the causes of the progressive impoverishment of coral reefs as the seas are warmed and made increasingly acid by the continued infusion of carbon dioxide from the atmospheric accumulation.[7] The reefs are unquestionably hot spots of biodiversity, now rapidly collapsing globally.

Biodiversity became the criterion for establishing an interest in conserving places, whether on land or in water. Hot spots of biodiversity were identified, and parks and reserves were established to preserve such sites. The emphasis on local parks seemed appropriate, as did attempts to connect parks with corridors left in their natural state for the benefit of migratory and otherwise wide-ranging animals. But again, the emphasis was typically narrow and specific. The parks, absolutely necessary and appropriate, all too frequently appeared to mollify the conservation constituency while leaving the rest of the world open to business as usual. Business as usual was in fact destroying all life on earth while scientists and conservationists quietly lived in their own intellectual ghetto, "successful," winning some battles but losing the war. I recall well a conversation with Russell Train, long after he had taken on the presidency of the World Wildlife Fund, in which he agreed that he had never thought we conservationists were doing more than slowing the destruction of nature. At that time and ever since I have thought that objective inadequate. I am sure he was in fundamental agreement and shared the frustration.

To be successful conservation had to become a prime purpose of government, a core effort. The objective had to shift from simply setting aside land or water in parks and reserves, or forbidding through the ESA this or that new development, to restoring and preserving the essential biotic character, the core natural communities of each region. The hazard was and remains erosion through chronic disturbance. But too many of the practitioners were preoccupied, trapped by their own success in advertising the Red List and the losses of eye-catching species. They were caught in their own attractively compelling arguments that "extinction is forever," and in developing the prestige and reach of their own not-for-profit organizations, which at best remained on the periphery of political power. Doors to the halls of political power were open to corporate exploiters, who knew their business and turned to government to favor growth and profits. Conservation, if it were to be successful, had to become competitive with corporate interests, not only in establishing parks and reserves, but also especially in showing that human welfare and the human future depend on preserving the continua of species in natural communities that are the operating systems of the biosphere.

Those insights are not simple. While the technical arguments advanced in the interests of preserving biodiversity were appropriate, and establishing parks and reserves remains honorable and essential, the focus on species was and remains inadequate to the point of being misleading. Any species exists as a population of plants or animals with an array of mutual dependencies in one or more natural communities. The population has a geographic range, large or small. Within that range, the genetic composition of the species varies. Each individual that survives has a genetic constitution appropriate for survival in that place and, quite possibly, not in another place even within the total range of the larger population of the species. The individual is a unique genetic strain, an ecotype. A hot spot of biodiversity may be fascinating for its diversity of species, but "saving" it is saving artifacts unless major efforts are under way to see that there are significant other related communities preserved, and the physical, chemical, and biotic integrity of the region is preserved as well.

Meanwhile, the juggernaut of economic and industrial development rolls over more and more of the residual natural communities and pushes the preservation of biodiversity to the margins. The urgency of preserving natural communities as the maintenance system of the human environment has been lost in the rhetoric of biodiversity, left in the hands of the specialists. Some have found a useful shorthand in framing "conservation" as the maintenance of biotic diversity, yet the concept offered no implement, no tool in science and no model, no example of success, and no compellingly clear cost of failure. It was a concept, attractive but ephemeral to most of the public, and not easily defined, measured, or conserved in practical, specific, understandable terms. Substantial efforts such as the proposal for a National Biological Survey were made to correct those deficiencies, but they proved feeble in the face of economic interests.

* * *

In many cases the establishment of reserves and even rules have been far too late, even to provide a moment's respite from the rasp of industrial expansion and exploitation. The examples are legion and part of all our lives. I recall, for instance, as a boy visiting family friends who lived on T-Wharf on the Boston waterfront. The throng of commercial fishing boats that crowded that large dock several vessels deep enthralled me. These were small, sturdy, diesel-powered wooden craft rigged with trawls that were put over the side and drawn aboard amidships: "beam trawlers" they were called, and I was fascinated as any boy would be. They were replaced in later years by larger "stern trawlers," which continue to this day to work deeper waters offshore. Back then, on family visits to the coast of Maine, I could watch the trawlers working the inshore fishery, and hear the comments of the locals about how the trawlers were taking everything and there was now nothing left. Perhaps it was hyperbole that there was nothing left, but the cod have now been gone from the inshore waters for a long time. Protected zones where no fishing of the species is allowed have recently been established, but it may be too late.

Years later I came back to New England to live in the marine community of Woods Hole, a village on a southern peninsula of Cape Cod where Spencer Baird had settled to investigate fish and fisheries in the 1870s. Convinced that the fisheries were in peril, Baird persuaded Congress to establish the US Fisheries Commission and became the organization's first commissioner, serving without pay. He was witnessing the late nineteenth-century intensification of the fisheries, big traps along the shore, the shift from wind to steam power, and the advent of trawlers that worked the inshore stocks of herring, mackerel, and cod along the New England coast.

Because cod grew to large size and could be "flaked," dried, salted, and traded far and wide, they were especially valuable and their abundance was legendary.[8] Early settlers on Cape Cod described them as so plentiful that cod could be scooped from the water in a bucket. The inshore strains of those cod stocks, despite their original abundance, were fished out and have never returned. Such populations, experience suggests, were ecotypes, genetically attuned to the inshore environments, temperatures, depths, currents, and food, and able to survive the assaults of all predators—except for the trawls. One might think the populations from the deeper waters would replace the stocks. Perhaps, in time, but those populations, too, are ecotypes, fixed to their special places and roles in the world.[9] The offshore populations are under pressure from the diesel-powered stern trawlers, which are larger, more efficient in their catch, far more powerful than their predecessors, and able to stay out for days or more if necessary, hauling larger trawls. Conservation? Biodiversity? Great thoughts, outside the discussion, forced out by the cries of fishermen as stocks declined and fishermen found themselves not only competing for diminished stocks but also at odds with efforts to reduce catches to preserve a residual community of cod capable of recovery from the harvest.

So, too, the once-huge cod stocks of the Grand Banks of Newfoundland were destroyed by overfishing and have not recovered despite many years of protection—and hope. Again, it was clearly the responsibility of government to protect not just the species but also the entire community of plants and animals of the banks that supported, among other species, a huge population of cod. The public interest in a long-term harvest was lost as the Grand Banks of Newfoundland—once, with Georges Bank a hundred miles off the New England shore, the richest fishery in the world—were driven by relentless commercial exploitation into a dismal impoverishment, persistent despite decades of attention and warnings.

The major residual fishery along the coast of eastern New England and eastern Canada in the first years of the twenty-first century has been, not surprisingly, low in the food web. The scavenger populations of lobsters have thrived in the much-reduced structure of the marine ecosystem as the populations of their predators, including the cod, have been greatly reduced. It is a sad, sad story, missed by those ignorant of what was, and now firmly attached to lobsters on the coast of Maine and New Brunswick as a culinary attraction.

The solution was and remains a fierce, effective program for preserving the biotic structure of the oceanic waters. Preservation at this late stage of exploitation requires restoration, assuming it is possible. In either case, the ruthless harvests have to be avoided over large areas set aside to provide a matrix of structural biotic integrity. These reserves must reach into every region: the oceanic surface communities, the deep waters, coastal, oceanic, bays and reefs, and tropical and polar. And they must include explicit protection of the chemistry of all waters. It is a large order, but there is no other way if the ocean's biotic resources are to be preserved. A human population of seven billion and still growing cannot survive long without a closely measured, effective effort to rebuild and preserve the global marine and coastal ecosystems.

The structure and exploitation of terrestrial ecosystems is quite different, but the prevalence of specific ecotypes reigns there as well. Forests prevail in influence on the biosphere by their area and massive stature, carbon content and metabolism. The forests, too, exist in a world of subspecific genetic specialization in which each forest stand represents the local survivors of the relentless selection that is the universal circumstance of living systems. As populations are lost to harvests and impoverishment from intensified disturbances, so ecotypes among all the species of the forest are also affected, not soon to be restored.

* * *

The search for a better rationale for conservation than species preservation has brought suggestions that biodiversity is essential for "the public service functions of nature," the normal metabolic processes of natural ecosystems that keep the biosphere operating. The concept is sound. The integrity of biospheric functions requires all species, and we can describe those functions as having financial value, as they do. Soil and actively flowing streams commonly restore sullied water at no explicit public cost and great public advantage. Where water must be filtered mechanically, the filtration plants can be expensive. Such natural "services" have status and value in the world of economics and politics, and the purpose in assigning a value is to claim that status in the competition with other financial interests. There have been proposals that such services be treated as commodities and sold as "ecosystem service units" available to those who find themselves in need of credits to make up for having destroyed such services elsewhere.[10] The cap-and-trade programs take advantage of the value of certain efforts in conservation or waste management to trade excesses with needy parties, often with a financial consideration. The Regional Greenhouse Gas Initiative described in chapter 6 for reducing the total greenhouse gas emissions within the northeastern states of the United States makes a market in credits for renewable energy among power companies forced by governmental regulations to reduce the use of fossil fuels. The electric power generated by the solar panels on my roof is being sold as credit to fossil-fuel-burning corporations that have not yet changed their power sources sufficiently to comply with new regulations.

While there are circumstances where there is clear advantage in assigning values to natural public services and using those values to protect them against commercial destruction, the monetary approach fails when the public values rise toward infinity, as often happens. In that case governmental control is required.

We live in and exploit for all aspects of our sustenance an environment that is a biotic system. It is dependent for its continuity of function on the totality of its elements, its total genetic pool, not only those recognized as species, but also all of its ecotypes, guilds, and communities of plants and animals, on land and in the global waters. It is the totality of that life that is now threatened with systematic impoverishment—a process that is vividly before us as the biotic feedbacks of the climatic disruption take the potential for control of climate out of our hands.

The enemy is chronic disturbance—cumulative, relentless, irreversible, physical, chemical, and biotic disturbance that moves the world systematically down the scale from cod and haddock to the lobsters and shellfish, down from the forest to the fire-seared blueberry barrens of eastern Maine. Impoverishment proceeds, place by place, until the effects fuse, and the disturbance becomes global and feeds on itself.

The cure starts with recognition that the physical, chemical, and biotic integrity of the biosphere is at issue, and preserving it must be seen as a call to arms because civilization hangs in the balance.

Notes

1 The most serious smog event in London, possibly the worst until current problems in China, occurred in December 1952, when several days of temperature inversion held coal smoke close to the ground and caused the death of thousands. Similar problems emerged in Pittsburgh subsequently and on occasion in many other industrial regions. See Devra Davis, *When Smoke Ran Like Water* (New York: Basic Books, 2003).

2 See various authors in E. O. Wilson, ed., *Biodiversity* (Washington, DC: National Academy Press, 1988). See also Norman Myers, "The Biodiversity Challenge: Expanded Hot-Spots Analysis," *Environmentalist* 10, no. 4 (1990): 243–256.

3 Richard O. Bierregaard Jr., Claude Gascon, Thomas E. Lovejoy, and Rita Mesquita, eds., *Lessons from Amazonia: The Ecology and Conservation of a Fragmented Forest* (New Haven, CT: Yale University Press, 2001). See also Jeff Tollefson, "Splinters of the Amazon," *Nature* 496 (2013): 286–289.

4 This topic has been examined in detail in the past from various angles including especially the obvious relationship between the number of species and size of the area considered—the species/area curve. Such analyses turn quickly to explorations of environmental gradients and the changes in communities associated with them. See, for instance, Robert H. Whittaker, *Communities and Ecosystems*, 2nd ed. (New York: Macmillan, 1975). More recently, various experimental approaches have looked at the relationships with diversity in a practical context. See David Tilman and John Downing, "Biodiversity and Stability in Grasslands," *Nature* 6461 (1994): 363–365; David Tilman, "Competition and Biodiversity in Spatially Structured Habitats," *Ecology* 75 (1994): 2–16.

5 National Research Council, *Biological Survey for the Nation* (Washington, DC: National Academy Press, 1993).

6 Endangered Species Act of 1973 (16 U.S.C. 1531–1544, 87 Stat. 884), as amended—Public Law 93–205, approved December 28, 1973, repealed the Endangered Species Conservation Act of December 5, 1969 (P.L. 91–135, 83 Stat. 275). The 1969 act had amended the Endangered Species Preservation Act of October 15, 1966 (P.L. 89–669, 80 Stat. 926). The 1973 act implemented the Convention on International Trade in Endangered Species of Wild Fauna and Flora (T.I.A.S. 8249), signed by the United States on March 3, 1973, and the Convention on Nature Protection and Wildlife Preservation in the Western Hemisphere (50 Stat. 1354), signed by the United States on October 12, 1940. See also Bruce Babbitt, *Cities in the Wilderness: A New Vision of Land Use in America* (New York: Island Press, 2005) 60, 74–75. Babbitt, as secretary of the interior, recognized the unique power of

the ESA in giving government powers over the management of public and private land and water in the public interest.

7 Elizabeth Kolbert, *The Sixth Extinction* (New York: Henry Holt, 2014).

8 Mark Kurlansky, *Cod: A Biography of a Fish That Changed the World* (Boston: Addison-Wesley, 1997).

9 I know of no direct study of these populations proving them distinct. I make the suggestion on the basis of numerous analogous studies of other plant and animal populations. We draw on such experience regularly in managing biotic resources. Foresters, for example, pay close attention to "provenance" in selecting seedlings for plantations. They want seedlings from populations that thrive in similar climates and soils. Virtually all populations are ecotypes, some more narrowly restricted than others. The inshore environment is substantially different from the deeper offshore waters, and the life cycles of populations of cod on the two sites were undoubtedly different. The fact that the inshore population remains low to nonexistent suggests little or no exchange between the two.

10 Gretchen G. Daily and Katherine Ellison, *The New Economy of Nature: The Quest to Make Conservation Profitable* (Washington, DC: Island Press, 2007). See also Paul Hawken, Amory Lovins, and L. Hunter Lovins, *Natural Capitalism: The Next Industrial Revolution* (London: Earthscan, 1999). Discussion adapted from George M. Woodwell, "The Biodiversity Blunder," *BioScience* 60, no. 11 (2010): 870–871.

Discussion Questions

1   What does Woodwell argue is the main reason for ignoring human impacts on the living world in the 20th century?

2   What does the Red List of the IUCN publicize?

3   How can you explain the near-elimination of the great cod fisheries of New England (even at Cape Cod), but a thriving lobster fishery?

4   What is the primary "tool" available at the U. S. federal level for attempting to protect threatened species?

5   The chapter author, George Woodwell, is an esteemed ecologist and conservationist, but he has a fundamental problem with the approach of trying to protect individual species. a) What does he think is more important? b) What interferes with achieving it? c) What ultimately "hangs in the balance" of protecting it?

PART II

CLIMATE AND ITS DISRUPTION

CHAPTER 3

The Scientific Consensus on Climate Change

How Do We Know We're Not Wrong?

Naomi Oreskes

In December 2004, *Discover* magazine ran an article on the top science stories of the year. One of these was climate change, and the story was the emergence of a scientific consensus over the reality of global warming. *National Geographic* similarly declared 2004 the year that global warming "got respect" (Roach 2004).

Many scientists felt that respect was overdue. As early as 1995, the Intergovernmental Panel on Climate Change (IPCC) had concluded that "the balance of evidence" supported the conclusion that humans were having an impact on the global climate (Houghton et al. 1995). By 2007, the IPCC's Fourth Assessment Report found a stronger voice, declaring warming "unequivocal" and noting it is "extremely unlikely that the global climate changes of the past fifty years can be explained without invoking human activities" (Alley et al. 2007). Prominent scientists and major scientific organizations have all ratified the IPCC conclusion (Oreskes 2004). Today, all but a tiny handful of climate scientists are convinced that earth's climate is heating up and that human activities are a primary driving cause (Doran and Zimmerman 2009; Anderegg et al. 2010).

Yet, a decade later, Americans continue to wonder. A 2006 poll reported in *Time* magazine found that only just over half (56 percent) of Americans thought average global temperatures had risen—despite the fact that virtually all climate scientists think they have (The Royal Society 2005). Since 2006, public opinion has wavered—influenced by short-term fluctuations in weather, as well as by political and cultural events whose relationship to climate change is indirect at best (Leiserowitz et al. 2012, and sources cited). But one thing that has remained consistent is a gap between the virtually unanimous opinion of scientists that man-made

Naomi Oreskes, Selections from "The Scientific Consensus on Climate Change: How Do We Know We're Not Wrong?," *Climate Change: What It Means for Us, Our Children, and Our Grandchildren*, ed. Joseph F. C. DiMento and Pamela Doughman, pp. 105-128, 135-148. Copyright © 2014 by MIT Press. Reprinted with permission.

climate change is underway and the continued doubts of a significant proportion of the American people (Leiserowitz et al. 2012; see also Borick et al. 2011). Moreover, as Jon Krosnick and his colleagues have stressed, while the scientific community has for some time believed the evidence for climate change "justifies substantial public concern," the public has not broadly shared that view (Krosnick et al. 2006; see also Lorenzoni and Pidgeon 2006).

This book addresses the scientific study of climate change and its effects. Its title draws our attention, in particular, to what climate change will mean for our children and grandchildren. By definition predictions are uncertain, and people may wonder why we should spend time, effort, and money addressing a problem that may not affect us for years or decades to come. Some people have gone further, suggesting that it would be foolish to spend time and money addressing a problem that might not actually even exist. After all, how do we really know?

This chapter addresses that issue: how *do* we really know? Put another way, even if there is a scientific consensus, how do we know it's not wrong? If the history of science teaches anything, it is humility. There are numerous historical examples of expert opinion that turned out to be wrong. At the start of the twentieth century, Max Planck was advised not to go into physics because all the important questions had been answered, medical doctors prescribed arsenic for stomach ailments, and geophysicists were confident that continents did not drift. In any scientific community there are individuals who depart from generally accepted views, and occasionally they turn out to be right. At present, there is a scientific consensus that climate change is underway, and that consensus has been stable for more than a decade. But how do we know it's not wrong?

The Scientific Consensus on Climate Change

Let's start with a simple question: what is the scientific consensus on climate change, and how do we know it exists? Scientists do not vote on contested issues, and most scientific questions are far too complex to be answered by a simple yes or no response. So how does anyone know what scientists think about global warming?

Scientists glean their colleagues' conclusions by reading their results in published scientific literature, listening to presentations at scientific conferences, and discussing data and ideas in the hallways of conference centers, university departments, research institutes, and government agencies. For outsiders, this information is difficult to access: scientific papers and conferences are written by experts, for experts, and are difficult for outsiders to understand.

Climate science is a little different. Because of the political importance of the topic, scientists have been motivated and asked to explain their research results in accessible ways, and explicit statements of the state of scientific knowledge are easy to find.

An obvious place to start is the Intergovernmental Panel on Climate Change (IPCC), already discussed in previous chapters. Created in 1988 by the World Meteorological Organization and the United Nations Environment Programme, the IPCC evaluates the state of climate science as a basis for informed policy action, primarily using peer-reviewed and published scientific litera-ture (IPCC 2013a). The IPCC has issued several assessments. In 2001, the IPCC had already stated unequivocally the consensus scientific opinion that earth's climate is being affected by human activities. This view is expressed throughout the report, but perhaps the clearest statement is this: "Human activities ... are modifying the concentration of atmospheric constituents ... that absorb or scatter radiant energy. ... Most of the observed warming over the last 50 years is likely to have been due to the increase in greenhouse gas concentrations" (McCarthy et al. 2001, 21). The 2007 IPCC report amends this to "very likely" (Alley et al. 2007). And the 2013 report added greater specificity, concluding, "It is *extremely likely* [greater than 95 percent confidence] that more than half of the observed increase in global average surface temperature from 1951 to 2010 was caused by the anthropogenic increase in greenhouse gas concentrations and other anthropogenic forcings together" (emphasis in original; IPCC 2013b, SPM-12).

From a historical perspective, the IPCC is a somewhat unusual scientific organization: it was created not to discover new knowledge but to compile and assess existing knowledge on a politically sensitive and economically significant issue. Perhaps its conclusions have been skewed by these extra-scientific concerns, but the IPCC is by no means alone it its conclusions; its results have been repeatedly ratified by other scientific organizations.

All of the major scientific bodies in the United States whose membership's expertise bears directly on the matter have issued reports or statements that confirm the IPCC conclusion. One is the National Academy of Sciences Committee on the Science of Climate Change report *Climate Change Science: An Analysis of Some Key Questions* (2001), which originated from a White House request. Here is how it opens: "Greenhouse gases are accumulating in Earth's atmosphere as a result of human activities, causing surface air temperatures and subsurface ocean temperatures to rise" (National Academy of Sciences Committee on the Science of Climate Change 2001, 1). The report explicitly addresses whether the IPCC assessment is a fair summary of professional scientific thinking, and answers yes: "The IPCC's conclusion that most of the observed warming of the last 50 years is likely to have been due to the increase in greenhouse gas concentrations accurately reflects the current thinking of the scientific community on this issue" (3).

Other US scientific groups agree. In February 2003, the American Meteorological Society adopted the following statement on climate change: "There is now clear evidence that the mean annual temperature at the Earth's surface, averaged over the entire globe, has been increasing in the past 200 years. There is also clear evidence that the abundance of greenhouse gases has increased over the same period. ... Because human activities are contributing to climate change, we have a collective responsibility to develop and undertake carefully considered response actions" (American Meteorological Society 2003). So too says the American Geophysical Union: "Scientific evidence strongly indicates that natural influences cannot explain the rapid increase in global near-surface temperatures observed during the second half of the 20th century" (American Geophysical Union Council 2003/2007). Likewise the American Association for the Advancement of Science: "The world is warming up. Average temperatures are half a degree centigrade higher than a century ago. ... Pollution from 'greenhouse gases' such as carbon dioxide (CO_2) and methane is at least partly to blame" (Harrison and Pearce 2000). In short, these groups all affirm that global warming is real and substantially attributable to human activities. In 2010, the National Academy of Sciences summarized, "Climate change is occurring, is caused largely by human activities, and poses significant risks for—and in many cases is already affecting—a broad range of human and natural systems" (3).

If we extend our purview beyond the United States, we find this conclusion further reinforced. In 2005, the Royal Society of the United Kingdom, one of the world's oldest and most respected scientific societies, issued a "Guide to Facts and Fictions about Climate Change," debunking various myths asserting that climate change is not occurring, that it is not caused by human activities, that observed changes are within the range of natural variability, that CO_2 is too trivial to matter, that climate models are unreliable, and that the IPCC is biased and does not fairly represent the scientific uncertainties.

The report takes pains to underscore the scientific authority of the IPCC, noting "the IPCC is the world's leading authority on climate change and its impacts" and that its work is backed by the worldwide scientific community.[1] This point was further underscored in 2007, when the National Academies of thirteen countries (the G8 plus another five) issued a joint statement calling attention to the problem of anthropogenic climate change and urging a rapid transition to a low-carbon society (Joint Science Academies 2008).

One website dedicated to evaluating the scientific consensus on climate change counts twenty-seven scientific societies that have formally endorsed the conclusion that "most of the global warming in recent decades can be attributed to human activities"—those just mentioned in North America, Europe, and Australia—as well as thirteen National Academies in Africa.[2] If we were to do a comprehensive count of scientific societies in Asia, Africa, and South America, the figure would no doubt be still higher.

Consensus reports and statements are drafted through a careful process involving many opportunities for comment, criticism, and revision, so it is unlikely that they would diverge greatly from the opinions of the societies' members. Nevertheless, it could be the case that they downplay dissenting opinions.[3]

One way to test that hypothesis is by analyzing the contents of published scientific papers, which contain the views that are considered sufficiently supported by evidence to merit publication in expert journals. After all, any one can *say* anything, but not anyone can get research results published in a refereed journal.[4] Papers published in scientific journals must pass the scrutiny of critical expert colleagues. They must be supported by sufficient evidence to convince others who know the subject well. So one must turn to the scientific literature to be certain of what scientists really think.

Before the twentieth century, this would have been a trivial task. The number of scientists directly involved in any given debate was usually small. A handful—a dozen, perhaps a hundred, at most—participated, in part because the total number of scientists in the world was very small

(Price 1986). Moreover, because professional science was a limited activity, many scientists used language that was accessible to scientists in other disciplines as well as to serious amateurs. It was relatively easy for an educated person in the nineteenth or early twentieth century to read a scientific book or paper and understand what the scientist was trying to say. One did not have to be a scientist to read *The Principles of Geology* or *The Origin of Species*.

Our contemporary world is different. Today, hundreds of thousands of scientists publish over a million scientific papers each year.[5] The American Geophysical Union has 50,000 members in 135 countries, and the American Meteorological Society has 14,000 members. The IPCC reports involved the participation of many hundreds of scientists from scores of countries (Houghton, Jenkins, and Ephraums 1990; Alley et al. 2007). No individual could possibly read all the scientific papers on a subject without making a full-time career of it.

Fortunately, the growth of science has been accompanied by a corresponding growth of tools to manage scientific information. One of the most important of these is the database of the Institute for Scientific Information (ISI). In its Web of Science, the ISI indexes all papers published in refereed scientific journals every year—over 8,500 journals. Using a key word or phrase, one can sample the scientific literature on any subject and get an unbiased view of the state of knowledge.

Figure 3.1 shows the results of an analysis of 928 abstracts, published in refereed journals during the period 1993 to 2003, that I completed in 2004 using the Web of Science database to evaluate the state of scientific debate at that time.[6]

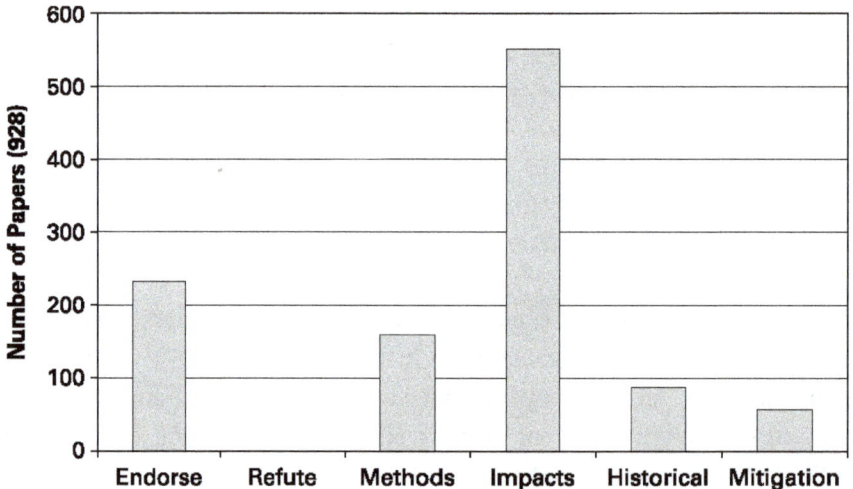

Figure 3.1 A Web of Science analysis of 928 abstracts using the keywords "global climate change." No papers in the sample provided scientific data or theoretical arguments to refute the consensus position on the reality of global climate change.

After a first reading to determine appropriate categories of analysis, the papers were divided as follows: (1) those explicitly endorsing the consensus position, (2) those explicitly refuting the consensus position, (3) those discussing methods and techniques for measuring, monitoring, or predicting climate change, (4) those discussing potential, or documenting actual, impacts of climate change, (5) those dealing with paleoclimate change, and (6) those proposing mitigation strategies. How many fell into category 2? That is, how many of these papers presented evidence refuting the statement, "Global climate change is occurring, and human activities are at least part of the reason why"? The answer is remarkable: none.

A few comments are in order. First, it is often challenging to determine exactly what the authors of a paper do think about global climate change. This is a consequence of experts writing for experts: many elements are implicit. If a conclusion is widely accepted, then it is not necessary to reiterate it within the context of expert discussion. Scientists generally focus their discussions on questions that are still disputed or unanswered rather than on matters about which everyone agrees.

This is clearly the case with the largest portion of the papers examined (approximately half of the total)—those dealing with the impacts of climate change. The authors evidently accept the premise that climate change is real and want to track, evaluate, and understand its consequences. Nevertheless, such consequences could, at least in principle, be the results of natural variability rather than human activities. Strikingly, none of the papers used that possibility to argue against the consensus position.

Roughly 15 percent of the papers dealt with methods, and slightly less than 10 percent dealt with paleoclimate change. The most notable trend in the data is the recent increase in such papers; concerns about global climate change have given a boost to research in paleoclimatology and to the development of methods for measuring and evaluating global temperature and climate. Such papers are essentially neutral with respect to the reality of current anthropogenic change: developing better methods and understanding historic climate change are important tools for evaluating current effects, but they do not commit their authors to any particular opinion about those effects. Perhaps some of these authors are in fact skeptical of the current consensus, and this could be a motivation to work on a better understanding of the natural climate variability of the past. But again, none of the papers used that motivation to argue openly against the consensus, and it would be illogical if they did because a skeptical motivation does not constitute scientific evidence. Finally, approximately 20 percent of the papers explicitly endorsed the consensus position, and an additional 5 percent proposed mitigation strategies. In short, by 2003, the basic reality of anthropogenic global climate change was no longer a subject of scientific debate.[7]

Some readers were surprised by this result and questioned the reliability of a study that failed to find arguments against the consensus position when such arguments clearly existed. After all, anyone who watched Fox news or MSNBC or trolled the Internet knew that there was an enormous debate about climate change, right? Well, no.

First, let's make clear what the scientific consensus is. It is over the reality of human-induced climate change. Scientists predicted a long time ago that increasing greenhouse gas emissions could change the climate, and now there is overwhelming evidence that it *is* changing the climate. These changes are *in addition* to natural variability. Therefore, when contrarians try to shift the focus of attention to natural climate variability, they are misrepresenting the situation. No one denies the fact of natural variability, but natural variability alone does not explain what we are now experiencing. Scientists have also documented that many of the changes that are now occurring are deleterious to both human and nonhuman communities (Root et al. 2003; Arctic Council 2004/2005; Hoegh-Guldberg 2005; Parmesan 2006; Adger et al. 2007.) Because of global warming, sea level is rising, humans are losing their homes and hunting grounds, plants and animals are shifting their ranges and in some cases losing their habitats, and extreme weather events (particularly droughts and heat waves) are becoming more common and in some cases more extreme (Kolbert 2006; Flannery 2006; Adger et al. 2007; IPCC 2012).

Second, to say that man-made global warming is underway is not the same as agreeing about what will happen in the future. Much of the continuing debate in the scientific community involves the likely rate of future change. A good analogy is evolution. In the early twentieth century, paleontologist George Gaylord Simpson introduced the concepts of "tempo and mode" to describe questions about the manner of evolution—how fast and in what manner evolution proceeded. Biologists by the mid-twentieth century agreed about the reality of evolution, but there were extensive debates about its tempo and mode. So it is now with climate change. Nearly all professional climate scientists agree that human-induced climate change is underway, but debate continues on tempo and mode.

Third, there is the question of what kind of dissent still exists. My analysis of the published literature was done by sampling published papers, using a keyword phrase that was intended to be fair, accurate, and neutral: "global climate change" (as opposed to, for example, "global warming," which might be viewed as biased). The *total* number of scientific papers published over that ten-year period having anything at all to do with climate change was over ten thousand; it is likely that some of the authors of the unsampled papers expressed skeptical or dissenting views. But given that the sample turned up no dissenting papers at all, professional dissension must have been very limited. Recent work has supported this conclusion, showing that 97–98 percent of professional climate scientists affirm the reality of anthropogenic climate change as outlined by the IPCC (Anderegg et al. 2010; see also Cook et al. 2013). This also affirms the conclusions of Max and Jules Boykoff (2004, see also Freudenburg and Muselli 2010; Boykoff 2011) that the mass media have given air and print space to a handful of dissenters to a degree that is greatly disproportionate with their representation in the scientific community. News articles on climate change, for example, may quote two mainstream scientists and one dissenter, where an accurate reflection of the state of the science would be to quote 30 or 40 mainstream scientists for every dissenter. (On television and radio the situation is even worse, where a debate is set up between one mainstream scientist and one dissenter, as if the actual distribution of views in the scientific community were fifty-fifty.) There are climate scientists who actively do research in the field but disagree with the consensus position, but their number is very, very small. This is not to say that there are not a significant number of *contrarians*, but to point out that the vast majority of them are not climate scientists.

In fact, most contrarians are not even scientists at all. Some, like the physicist Frederick Seitz (who for many years challenged the scientific evidence of the harms of tobacco along with the threat of climate change), were once scientific researchers but not in the field of climate science. (Seitz was a solid-state physicist.) Others, like Michael Crichton, who for many years was a prominent speaker on the contrarian lecture circuit, are novelists, actors, or others with access to the media, but no scientific credentials. What Seitz and Crichton had in common, along with most other contrarians, is that they did little or no new scientific research. They were not producing new evidence or new arguments to be judged by scientists in the halls of science. They were attacking the work of others, and doing so in the court of public opinion and in the mass media.

This latter point is crucial and merits underscoring: the vast majority of books, articles, and websites denying the reality of global warming do not pass the most basic test for what it takes to be counted as scientific—namely, being published in a peer-reviewed journal. Contrarian views have been published in books and pamphlets issued by politically motivated think tanks and widely spread across the Internet (Jacques et al. 2008), but so have views promoting the reality of UFOs or the claim that Lee Harvey Oswald was an agent of the Soviet Union.

Moreover, some contrarian arguments are frankly disingenuous, giving the impression of refuting the scientific consensus when their own data do no such thing. One example will illustrate the point. In 2001, Willie Soon, a physicist at the Harvard-Smithsonian Center for Astrophysics, with several colleagues published a paper entitled "Modeling Climatic Effects of Anthropogenic Carbon Dioxide Emissions: Unknowns and Uncertainties" (Soon et al. 2001). This paper has been widely cited by contrarians as an important example of a legitimate dissenting scientific view published in a peer-reviewed journal.[8] But the issue under discussion is how well models can predict the future—in other words, tempo and mode. The paper does not refute the consensus position, and the authors acknowledge so: "The purpose of [our] review of the deficiencies of climate model physics and the use of GCMs is to illuminate areas for improvement. Our review does not disprove a significant anthropogenic influence on global climate" (Soon et al. 2001, 259; see also Soon et al. 2002).

The authors needed to make this disclaimer because many contrarians do try to create the impression that arguments about tempo and mode undermine the whole picture of global climate change. But they don't. Indeed, one could reject all climate models and still accept the consensus position because models are only one part of the argument—one line of evidence among many.

Is there disagreement over the details of climate change? Yes. Are all the aspects of climate past and present well understood? No, but who has ever claimed that they were? Does climate science tell us what policy to pursue? Definitely not, but it does identify the problem, explain why it matters, and give society insights that can help to frame an efficacious policy response (e.g., Smith 2002; Oreskes, Smith, and Stainforth 2010).

So why does the public have the impression of disagreement among scientists? If the scientific community has forged a consensus, then why do so many Americans have the impression that there is serious scientific uncertainty about climate change?[9]

There are several reasons. First, it is important to distinguish between scientific and political uncertainties. There are reasonable differences of opinion about how best to respond to climate change and even about how serious global warming is relative to other environmental and social issues. Some people have confused—or deliberately conflated—these two issues. Scientists are in agreement about the reality of global climate change, but this does not tell us what to do about it.

Second, climate science involves prediction of future effects, which by definition are uncertain. It is important to distinguish among what is known to be happening now, what is likely to happen based on current scientific understanding, and what might happen in a worst-case scenario. This is not always easy to do, and scientists have not always been effective in making these distinctions. Uncertainties about the future are easily conflated with uncertainties about the current state of scientific knowledge.

Third, scientists have evidently not managed to explain well enough their arguments and evidence beyond their own expert communities. The scientific societies have tried to communicate to the public through their statements and reports on climate change, but what average citizen knows that the American Meteorological Society even exists or visits its home page to look for its climate-change statement?

There is also a deeper problem. Scientists are finely honed specialists trained to create new knowledge, but they generally have limited training in how to communicate to broad audiences and even less in how to defend scientific work against determined and well-financed contrarians (Moser and Dilling 2004, 2007; Hassol 2008; Somerville and Hassol 2011). Moreover, until recently, most scientists have not been particularly anxious to take the time to communicate their message broadly. Most scientists consider their "real" work to be the production of knowledge, not its dissemination, and often view these two activities as mutually exclusive. Some even sneer at colleagues who communicate to broader audiences, dismissing them as "popularizers" (Olson 2009).

If scientists do jump into the fray on a politically contested issue, they may be accused of "politicizing" the science and compromising their objectivity.[10] This places scientists in a double bind: the demands of objectivity suggest that they should keep aloof from contested issues, but if they don't get involved, no one will know what an objective view of the matter looks like. Scientists' reluctance to present their results to broad audiences has left scientific knowledge open to misrepresentation, and recent events show that there are plenty of people ready and willing to misrepresent it.

It's no secret that politically motivated think tanks such as the American Enterprise Institute and the George C. Marshall Institute have been active for some time in trying to communicate a message that is at odds with the consensus scientific view (Gelbspan 1997, 2004; Mooney 2006; Jacques et al. 2008; Hoggan and Littlemore 2009; Oreskes and Conway 2010). These organizations have successfully garnered a great deal of media attention for the tiny number of scientists who disagree with the mainstream view and for nonscientists, like Crichton, who pronounce loudly on scientific issues.

This message of scientific uncertainty has been reinforced by the public relations campaigns of certain corporations with a large stake in the issue.[11] The most well-known example is ExxonMobil, which in 2000 and 2004 ran highly visible advertising campaigns on the op-ed page of the *New York Times*. Its carefully worded advertisements—written and for-matted to look like newspaper columns and called op-ed pieces by ExxonMobil—suggested that climate science was far too uncertain to warrant action on it.[12] The claims made in these advertisements were not literally untrue, but they were, arguably, very misleading. In 2011 and 2012, ExxonMobil expressed concern about climate change in corporate reports but continued to argue for delay in other venues (Union of Concerned Scientists 2012). Our scientists have long ago concluded that existing research warrants that decisions and policies be made *today*.[13]

In any scientific debate, past or present, one can always find intellectual outliers who di-verge from the consensus view. Even after plate tectonics was resoundingly accepted by earth scientists in the late 1960s, a handful of persistent resisters clung to the older views, and some idiosyncratics held to alternative theoretical positions, such as earth expansion. Some of these men were otherwise respected scientists, including Sir Harold Jefferys, one of Britain's leading geophysicists, and Gordon J. F. MacDonald, a one-time science adviser to Presidents Lyndon Johnson and Richard Nixon. Both these men rejected plate tectonics until their dying day, which for MacDonald was in 2002. Does that mean that scientists should reject plate tectonics, that disaster-preparedness campaigns should not use plate-tectonic theory to estimate regional earthquake risk, or that schoolteachers should give equal time in science classrooms to the theory of earth expansion? Of course not. That would be silly and a waste of time. In the case of earthquake preparedness, it would be dangerous as well.

No scientific conclusion can ever be proven, and new evidence may lead scientists to change their views, but it is no more a "belief" to say that earth is heating up than to say that continents move, that germs cause disease, that DNA carries hereditary information, that HIV causes AIDS, and that some synthetic organic chemicals can disrupt endocrine function. You can always find someone, somewhere, to disagree, but these conclusions represent our best current understand-ing and therefore our best basis for reasoned action (Oreskes 2004).

How Do We Know We're Not Wrong?

Might the consensus on climate change be wrong? Yes, it might be, and if scientific research continues, it is almost certain that some aspects of the current understanding will be modified, perhaps in significant ways. This possibility can't be denied. The relevant question for us as citizens is not whether this scientific consensus *might* be mistaken but rather whether there is any reason to think that it *is* mistaken.

How can outsiders evaluate the robustness of any particular body of scientific knowledge? Many people expect a simple answer to this question. Perhaps they were taught in school that scientists follow "the scientific method" to get correct answers, and they have heard some climate-change deniers suggesting that climate scientists do not follow the scientific method (because they rely on models, rather than laboratory experiments) so their results are suspect. These views are wrong.

Contrary to popular opinion, there is no scientific method (singular). Despite heroic efforts by historians, philosophers, and sociologists, there is no generally agreed-upon answer as to what the methods and standards of science are (or even what they should be). There is no methodological litmus test for scientific reliability and no single method that guarantees valid conclusions that will stand up to all future scrutiny.

A positive way of saying this is that scientists have used a variety of methods and standards to good effect and that philosophers have proposed various helpful criteria for evaluating the methods used by scientists. None is a magic bullet, but each can be useful for thinking about what makes scientific information a reliable basis for action.[14] So we can pose the question: how does current scientific knowledge about climate stand up to these diverse models of scientific reliability?

The Inductive and Deductive Models of Science

The most widely cited models for understanding scientific reasoning are induction and deduction. *Induction* is the process of generalizing from specific examples. If I see 100 swans and they are all white, I might conclude that all swans are white. If I saw 1,000 white swans or 10,000, I would surely think that *all* swans were white, yet a black one might still be lurking somewhere. As David Hume famously put it, even though the sun has risen thousands of times before, we cannot *prove* that it will rise again tomorrow.

Nevertheless, common sense tells us that the sun will rise again tomorrow, even if we can't logically prove that it's so. Common sense similarly tells us that if we had seen ten thousand white swans, then our conclusion that all swans were white would be more robust than if we had seen only ten. Other things being equal, the more we know about a subject, and the longer we have studied it, the more likely our conclusions about it are to be true.

How does climate science stand up to the inductive model? Does climate science rest on a strong inductive base? Yes. Humans have been making temperature records consistently for over 150 years, and nearly all scientists who have looked carefully at these records see an overall temperature increase since the Industrial Revolution (Houghton, Jenkins, and Ephraums 1990; Bruce et al. 1996; Watson et al. 1996; McCarthy et al. 2001; Houghton et al. 2001; Metz et al. 2001; Watson 2001; Weart 2003). According to the Climate Change 2007 Synthesis Report of the IPCC's Fourth Assessment Report, the temperature rise over the 100-year period from 1906 to 2005 was 0.74°C (0.56 to 0.92°C) with a confidence interval of 90 percent (2007a, 27–30). The IPCC's Fifth Assessment Report said temperature for the end of the twenty-first century is "*likely* to exceed 1.5 degrees C relative to 1850 to 1900" for all but one scenario included in the analysis (emphasis in original; IPCC 2013b, SPM-15).

How reliable are the early records? And how do you average data to be representative of the globe as a whole, when most of the early data come from only a few places, generally in Europe? Scientists have spent quite a bit of time addressing these questions; most have satisfied themselves that the empirical signal is clear (Edwards, 2010). Even if scientists doubted the older records, the more recent data show a strong increase in temperatures over the past thirty to forty years, just when the amount of carbon dioxide and other greenhouses gases in the atmosphere was growing dramatically (McCarthy et al. 2001; Houghton et al. 2001; Metz et al. 2001; Watson 2001). Recently, an independent assessment by the Berkeley Earth Surface Temperature group found that over the past fifty years the land surface warmed by 0.91°C, a result that confirms the prior work by NASA, the National Oceanic and Atmospheric Administration, and the Hadley Centre (Rohde et al. 2013). The Berkeley group has also reviewed the question of the "heat island effect"—the possible exaggeration of the warming effect due to the location of weather stations in urban areas, which are warmer than rural ones because of buildings, concrete, automobiles, and the like—a potential source of error much emphasized by some contrarians (Wickham et al. 2013), and finds that the observed warming cannot be explained away as an artifact of the heat island effect.

The Berkeley study received a good deal of media attention—arguably out of proportion to its scientific significance—because its spokesman, physicist Richard Muller, was previously a self-proclaimed skeptic, and because some of his funding came from the Koch Industries, a Fortune 500 company heavily involved in petroleum refining, oil and gas pipelines, and petrochemicals. (Both Koch brothers are political libertarians who are generally opposed to environmental regulation: David Koch ran in 1980 for Vice President on the Libertarian party ticket, and Charles Koch is one of the founders of the Cato Institute, which has played a large role in US climate change denial; see Oreskes and Conway 2010.) But despite a flurry of media attention, Richard Muller's late-stage conversion had little political, and even less scientific, impact because the conclusions from the instrumental records that he first questioned but then affirmed have been amply corroborated by other independent evidence from tree rings, ice cores, and coral reefs (IPCC 2007b, 438–439). A paper in 2002 by a team led by Jan Esper at the Swiss

Federal Research Center, for example, had already demonstrated that tree rings can provide a reliable, long-term record of temperature variability, one that largely (albeit not entirely) agrees with the instrumental records over the past 150 years (Esper, Cook, and Schweingruber 2002).[15]

Muller's reanalysis of existing temperature records raises the fundamental problem facing all inductive science: how many data are enough? If you have counted 10,000 white swans—or 100,000, or even 1,000,000—how do you know that a black swan does not exist elsewhere? And how do you know that the generalization you made from your observations is correct? After all, other generalizations could also be consistent with your observations.

The logical limitations of the inductive view of science have led some to argue that the core of scientific method is testing theories through logical deductions. *Deduction* is drawing logical inferences from a set of premises—the stock in trade of Sherlock Holmes. In science, deduction is generally presumed to work as part of what has come to be known as the *hypothetico-deductive model*—the model you will find in most textbooks that claim to teach the scientific method (sometimes also called the *deductive-nomological* model, referring to the idea that ultimately science seeks to develop not just hypotheses, but laws).

In this view, scientists develop hypotheses and then test them. Every hypothesis has logical consequences—deductions—and one can try to determine, primarily through experiment and observation, whether the deductions are correct. If they are, they support the hypothesis. If they are not, then the hypothesis must be revised or rejected. It's often considered especially good if the prediction is something that would otherwise be quite unexpected, because that would suggest it didn't just happen by chance.

The most famous example of successful deduction in the history of science is the case of Ignaz Semmelweis, who in the 1840s deduced the importance of hand washing to prevent the spread of infection (Gillispie 1975; Hempel 1965). Semmelweis had noticed that many women were dying of fever after giving birth at his Viennese hospital. Surprisingly, women who had their infants on the way to the hospital—seemingly under more adverse conditions—rarely died of fever. Nor did women who gave birth at another hospital clinic where they were attended by midwives. Not surprisingly, Semmelweis was troubled by this pattern, which seemed to suggest that it was more dangerous to give birth when attended by a doctor than by a midwife, and more dangerous to give birth in a hospital than in a horse-drawn carriage.

In 1847, a friend of Semmelweis, Jakob Kolletschka, cut his finger while doing an autopsy and soon died. Autopsy revealed a pathology very similar to the women who had died after childbirth; something in the cadaver had apparently caused his death. Semmelweis knew that many of the doctors at his clinic routinely went directly from conducting autopsies to attending births, but midwives did not perform autopsies. So he hypothesized that the doctors were carrying cadaveric material on their hands, which was infecting the women (and killed his friend). He deduced that if physicians washed their hands before attending the women, the infection rate would decline. Physicians did, and the infection rate declined, demonstrating the power of the hypothetico-deductive method.

How does climate science stand up to this standard? Have climate scientists made predictions that have come true? Absolutely. The most obvious is the fact of global warming itself. As already noted in previous chapters, scientific concern over the effects of increased atmospheric carbon dioxide is based on physics—the fact that carbon dioxide is a greenhouse gas, a fact that has been known since the mid-nineteenth century. In the early twentieth century, Swedish chemist Svante Arrhenius predicted that increasing carbon dioxide from the burning of fossil fuels would lead to global warming, and by midcentury, a number of other scientists, including G. S. Callendar, Roger Revelle, and Hans Suess, concluded that the effect might soon be noticeable, leading to sea level rise and other global changes (Fleming 1998; Weart 2003). In 1965, Revelle and his colleagues wrote: "By the year 2000, the increase in atmospheric CO_2 ... may be sufficient to produce measurable and perhaps marked change in climate, and will almost certainly cause significant changes in the temperature and other properties of the stratosphere" (Revelle 1965, 9). This prediction has come true (McCarthy et al. 2001; Houghton et al. 2001; Metz et al. 2001; Watson 2001).

Another prediction fits the category of something unusual that you might not even think of without the relevant theory. In 1980, climatologist Suki Manabe predicted that the effects of global warming would be strongest first in the polar regions. *Polar amplification* was not an induction from observations but a deduction from theoretical principles: the concept of ice albedo feedback. The reflectivity of a material is called its *albedo*. Ice has a high albedo, reflecting sunlight into space much more effectively than grass, dirt, or water. One reason polar regions are as cold as they are is that snow and ice are very effective in reflecting solar radiation back into space. But if the snow starts to melt and bare ground (or water) is exposed, this reflective effect diminishes. Less ice means less reflection, which means more solar heat is absorbed, leading to yet more melting in a feedback loop. So once warming begins, its effects accelerate; Manabe and his colleagues thus predicted that warming would be more pronounced in polar regions than in temperate ones. The Arctic Climate Impact Assessment concluded in 2004 that this prediction had come true (Manabe and Stouffer 1980, 1994; Holland and Bitz 2003; Arctic Council 2004/2005). [...]

Consilience of Evidence

Most philosophers and historians of science agree that there is no ironclad means by which to prove a scientific theory. But if science does not provide proof, then what is the purpose of induction, hypothesis testing, and falsification? Most would answer that, in various ways, these activities provide a warrant—or a justification—for our views. Do they?

An older view, which has come back into fashion of late, is that scientists look for consilience of evidence. *Consilience* means "coming together," and its use is generally credited to the English philosopher William Whewell, who defined it as the process by which sets of data—independently derived—coincided and came to be understood as explicable by the same theoretical account (Gillispie 1981; Wilson 1998). The idea is not so different from what happens in a legal case. To prove a defendant guilty beyond a reasonable doubt, a prosecutor must present a variety of evidence that holds together in a consistent story. The defense, in contrast, might need to show only that some element of the story is at odds with another to sow reasonable doubt in the minds of the jurors. In other words, scientists are more like lawyers than they might like to admit. They look for independent lines of evidence that hold together.

Do climate scientists have a consilience of evidence? Again the answer is yes. Instrumental records, tree rings, ice cores, borehole data, and coral reefs all point to the same conclusion: things are getting warmer overall. Keith Briffa and Timothy Osborn of the Climate Research Unit of the University of East Anglia compared the tree-ring analysis by Esper, Cook, and Schweingruber (2002) with six other reconstructions of global temperature between the years 1000 and 2000 (Briffa and Osborn 2002). All seven analyses agree: temperatures increased dramatically in the late twentieth century relative to the record of the previous millennium.

Inference to the Best Explanation

The various problems in trying to develop an account of how and why scientific knowledge is reliable have led some philosophers to conclude that the purpose of science is not proof, but explanation. Not just any explanation will do, however; the best explanation is the one that is consistent with the evidence (e.g., Lipton 1991). Certainly, it is possible that a malicious or mischievous deity placed fossils throughout the geological record to trick us into believing organic evolution—perhaps to test our faith?—but to a scientist this is not the best explanation because it invokes supernatural effects, and the supernatural is beyond the scope of scientific explanation. (It might not be the best explanation to a theologian, either, if that theologian was committed to heavenly benevolence.) Similarly, I might try to explain the drift of the continents through the theory of the expanding earth—as some scientists did in the 1950s—but this would not be the best explanation because it fails to explain why the earth has conspicuous zones of compression as well as tension. The philosopher of science Peter Lipton has put it this way: every set of facts has a diversity of possible explanations, but "we cannot infer something simply because it is a possible explanation. It must somehow be the best of competing explanations" (Lipton 2004, 56). Isaac Newton, in the *Principia Mathematica*, argued that our explanations must invoke causes that we know actually exist—so called *vera causae*. Invoking Martian hunting to explain the extinction of the dinosaurs would not be an inference to the best explanation, because we have no evidence that Martians exist, but invoking a meteorite can be, because large meteorites do.

Best is a term of judgment, so it doesn't entirely solve our problem, but it gets us thinking about what it means for a scientific explanation to be the best available—or even just a good one. It also invites us to ask the question, "Best for what purpose?" For philosophers, *best* generally means that an explanation is consistent with all the available evidence (not just selected portions of it), and that the explanation is consistent with other known laws of nature and other bodies of accepted evidence (and not in conflict with them). In other words, *best* can be judged in terms of the various criterion invoked by *all* the models of science discussed above: Is there an inductive basis? Does the theory pass deductive tests? Do the various elements of the theory fit with each other and with other established scientific information? And is the explanation potentially refutable and not invoking unknown, inexplicable, or supernatural causes?

Contrarians have tried to suggest that the climate effects we are experiencing are simply natural variability. Climate does vary, so this is a *possible* explanation. No one denies that. But is it the *best* explanation for what is happening now? Most climate scientists would say that it's not the best explanation. In fact, it's not even a good explanation—because it is inconsistent with much of what we know.

Should we believe that the global increase in atmospheric carbon dioxide has had a negligible effect, even though basic physics tells us it should be otherwise? Should we believe that the correlation between increased CO_2 and increased temperature is just a peculiar coincidence? If there were no theoretical reason to relate them, and if Arrhenius, Callendar, Suess, and Revelle had not predicted that all this would all happen, then one might well conclude that rising CO_2 and rising temperature were merely coincidental. But we have many reasons to believe that there is a causal connection and no good reason to believe that it is a coincidence. Indeed, the only reason we might think otherwise is to avoid committing to action: if this is just a natural cycle in which humans have played no role, then global warming might go away on its own in due course, and we would not have to spend money or be otherwise inconvenienced to remedy the problem.

And that sums things up. To deny that global warming is real is to deny that humans have become geological agents, changing the most basic physical processes of the earth, and therefore to deny that we bear responsibility for adverse changes that are taking place around us. For centuries, scientists thought that earth processes were so large and powerful that nothing we could do would change them. This was a basic tenet of geological science: that human chronologies were insignificant compared with the vastness of geological time; that human activities were insignificant compared with the force of geological processes. And once they were. But no more. There are now so many of us cutting down so many trees and burning so many billions of tons of fossil fuels that we have become geological agents. We have changed the chemistry of our atmosphere, causing sea level to rise, ice to melt, and climate to change. There is no reason to think otherwise. And, in my view, there is at this point in history no excuse for not taking action to prevent the very significant losses that are likely to ensue—indeed, losses that are already becoming evident.

Notes

1 For additional debunking of myths advanced by climate skeptics, see John Cook et al. (2013), "Skeptical Science: getting skeptical about global warming skepticism." http://www.skeptical-science.com/global-warming-scientific-consensus-intermediate.htm, accessed September 23, 2013.

2 Contrast this with the results of the Intergovernmental Panel on Climate Change's Third and Fourth Assessment Reports, which stated unequivocally that average global temperatures have risen (Houghton et al. 2001; Alley et al. 2007).

3 It should be acknowledged that in any area of human endeavor, leadership may diverge from the views of the led. For example, many Catholic priests endorse the idea that priests should be permitted to marry (Watkin 2004).

4 In recent years, climate-change deniers have increasingly turned to nonscientific literature as a way to promulgate views that are rejected by most scientists (see, for example, Deming 2005).

5 An e-mail inquiry to the Thomson Scientific Customer Technical Help Desk produced this reply: "We index the following number of papers in Science Citation Index—2004, 1,057,061 papers; 2003, 1,111,398 papers."

6 The analysis begins in 1993 because that is the first year for which the database consistently published abstracts. Some abstracts initially compiled were deleted from our analysis because the authors of those papers had put "global climate change" in their key words, but their papers were not actually on the subject.

7 This is consistent with the analysis of historian Spencer Weart, who concluded that scientists achieved consensus in 1995 (see Weart 2003).

8 In e-mails that I received after publishing my essay in *Science* (Oreskes 2004), this paper was frequently invoked. It did appear in the sample.

9 According to *Time* magazine, in 2006 a poll reported that, "64 percent of Americans think scientists disagree with one another about global warming" (*Time* 2006; ABC News/Time/Stanford Poll 2006).

10 Objectivity certainly can be compromised when scientists address charged issues. This is not an abstract concern. It has been demonstrated that scientists who accept research funds from the tobacco industry are much more likely to publish research results that deny or downplay the hazards of smoking than those who get their funds from the National Institutes of Health, the American Cancer Society, or other nonprofit agencies (Bero 2003). On the other hand, there is a large difference between accepting funds from a patron with a clearly vested interest in a particular epistemic outcome and simply trying one's best to communicate the results of one's research clearly and in plain English.

11 Some petroleum companies, such as BP and Shell, have made public efforts to acknowledge the reality of anthropogenic climate change and to refrain from participating in misinformation campaigns (see Browne 1997). Browne began his 1997 lecture by focusing on what he accepted as "two stark facts. The concentration of carbon dioxide in the atmosphere is rising, and the temperature of the Earth's surface is increasing." On the other hand, both BP and Shell were part of the Global Climate Coalition (see Gelbspan 1997, 2004), which promoted disinformation in the early to mid-1990s, during negotiations related to the UN Framework Convention on Climate Change and the Kyoto Protocol. Moreover, after an initial flurry of attention caused by Lord Browne's public statements, BP continued to develop its petroleum resources and only to put modest efforts into developing renewables and carbon sequestration technologies. For an analysis of diverse corporate responses, see Van den Hove et al. (2003).

12 An interesting development in 2003 was that Institutional Shareholders Services advised ExxonMobil shareholders to ask the company to explain its stance on climate-change issues and to divulge financial risks that could be associated with it (see *Planet Ark* 2003).

13 These efforts to generate an aura of uncertainty and disagreement have had an effect. This issue has been studied in detail by academic researchers (see, for example, Boykoff and Boykoff 2004).

14 *Reliable* is a term of judgment. By *reliable basis for action*, I mean that it will not lead us far astray in pursuing our goals, or if it does lead us astray, at least we will be able to look back and say honestly that we did the best we could given what we knew at the time.

15 For further discussion, see Esper, Frank, and Timonen et al. 2012; and Briffa, Melvin, Osborn et al. 2013.

References

ABC News/Time/Stanford Poll. 2006. March 9–14. http://www.pollingreport.com/enviro3.htm, accessed July 8, 2013.

Adger, Neil, Pramod Aggarwal, Shardul Agrawala, Joseph Alcamo, Abdelkader Allali, Oleg Anisimov, Nigel Arnell, et al. 2007. Climate change impacts, adaptation, and vulnerability: Summary for Policymakers. In *Climate Change 2007: Impacts, Adaptation and Vulnerability. Contribution of Working Group II to the Fourth Assessment Report of the Intergovernmental Panel on Climate Change*, edited by M. L. Parry, O. F. Canziani, J. P. Palutikof, P. J. van der Linden, and C. E. Hanson, 7–22. Cambridge, UK: Cambridge University Press.

Alley, Richard, Terje Berntsen, Nathaniel L. Bindoff, Zhenlin Chen, Amnat Chidthaisong, Pierre Friedlingstein, Jonathan M. Gregory, et al. 2007. IPCC 2007: Summary for Policymakers. In *Climate Change 2007: The Physical Science Basis. Contribution of Working Group I to the Fourth Assessment Report of the Intergovernmental Panel on Climate Change*, edited by S. Solomon, D. Qin, M. Manning, Z. Chen, M. Marquis, K. B. Averyt, M. Tignor, and H. L. Miller. New York: Cambridge University Press.

American Geophysical Union Council. 2003/2007. AGU Position Statement: Human Impacts of Climate. Adopted by Council December 2003. Revised and Reaffirmed December 2007. American Geophysical

Union, Washington, DC. http://www.agu.org/sci_pol/positions/climate_change2008.shtml, accessed July 8, 2013.

American Meteorological Society. 2003. Climate change research: Issues for the atmospheric and related sciences. *Bulletin of the American Meteorological Society* 84:508–515. http://www.ametsoc.org/policy/amsstatements_archive.html, accessed July 8, 2013.

Anderegg, W. R. L., James W. Pratt, Jacob Harold, and Stephen H. Schneider. 2010. Expert credibility in climate change. *Proceedings of the National Academy of Sciences* 107(27):12107–12109.

Arctic Council. 2004/2005. 2004 *Arctic climate impact assessment.* Arctic Council, Oslo, Norway. Cambridge, UK: Cambridge University Press. Available at http://www.amap.no/arctic-climate-impact-assessment-acia, accessed July 8, 2013.

Bero, L. 2003. Implications of the tobacco industry documents for public health and policy. *Annual Review of Public Health* 24:267–288.

Borick, Christopher P., Erick Lachapelle, and Barry Rabe. 2011. Climate compared: Public opinion on climate change in the United States and Canada. Brookings Institute. http://www.brookings.edu/research/papers/2011/04/climate-change-opinion, accessed July 8, 2013.

Boykoff, M. 2011. *Who Speaks for the Climate? Making Sense of Media Coverage of Climate Change.* Cambridge, UK: Cambridge University Press.

Boykoff, M. T., and J. M. Boykoff. 2004. Balance as bias: Global warming and the U.S. prestige press. *Global Environmental Change* 14:125–136.

Briffa, K. R., Thomas M. Melvin, Timothy J. Osborn, Rashit M. Hantemirov, Alexander V. Kirdyanov, Valeriy S. Mazepa, Stepan G. Shiyatov, et al. 2013. Reassessing the evidence for tree-growth and inferred temperature change during the Common Era in Yamalia, northwest Siberia. July 15. *Quaternary Science Reviews* 72:83–107.

Briffa, K. R., and T. J. Osborn. 2002. Blowing hot and cold. *Science* 295:2227–2228.

Browne, E. J. P. 1997. Climate change: The new agenda. Paper presented at Stanford University, May 19, Group Media and Publications, British Petroleum Company.

Bruce, James P., Hoesung Lee, and Erik F. Haites, eds. 1996. *Climate Change 1995: Economic and Social Dimensions of Climate Change. Intergovernmental Panel on Climate Change.* Cambridge, UK: Cambridge University Press.

Cook, J., D. Nuccitelli, S. A. Green, M. Richardson, B. Winkler, R. Painting, R. Way, et al. 2013. Quantifying the consensus on anthropogenic global warming in the scientific literature. *Environmental Research Letters* 8:024024. doi: 10.1088/1748–9326/8/2/024024.

Deming, David. 2005. How "consensus" on global warming is used to justify draconian reform. *Investor's Business Daily*, March 18, A16.

Doran, Peter T., and Maggie Kendall Zimmerman. 2009. Examining the scientific consensus on climate change. *Eos* 90(3):22–23.

Edwards, Paul. 2010. *A Vast Machine: Computer Models, Climate Data, and the Politics of Global Warming.* Cambridge, MA: MIT Press.

Esper, J., E. R. Cook, and F. H. Schweingruber. 2002. Low-frequency signals in long tree-ring chronologies for reconstructing past temperature variability. *Science* 295:2250–2253.

Esper, Jan, David C. Frank, Mauri Timonen, Eduardo Zorita, Rob J. S. Wilson, Jürg Luterbacher, Steffen Holzkämper, et al. 2012. Orbital forcing of tree-ring data. Letters. Published online July 8, 2012. doi: 10.1038. *Nature Climate Change* 1589:1–5.

Flannery, Tim. 2006. *The Weather Makers: How Man Is Changing the Climate and What It Means for Life on Earth.* New York: Atlantic Monthly Press.

Fleming, James Rodger. 1998. *Historical Perspectives on Climate Change*. New York: Oxford University Press.

Freudenburg, William R., and Violetta Muselli. 2010. Global warming estimates, media expectations, and the asymmetry of scientific challenge. *Global Environmental Change* 20(3):483–491.

Gelbspan, Ross. 1997. *The Heat Is On: The High Stakes Battle over Earth's Threatened Climate*. Reading, MA: Addison-Wesley.

Gelbspan, Ross. 2004. *Boiling Point: How Politicians, Big Oil and Coal, Journalists, and Activists are Fueling the Climate Crisis—And What We Can Do to Avert Disaster*. New York: Basic Books.

Gillispie, Charles C., ed. 1975. Semmelweis. In *Dictionary of Scientific Biography*, vol. 12. New York: Scribner.

Gillispie, Charles C., ed. 1981. *Dictionary of scientific biography*. Vol. 12. New York: Scribner.

Harrison, Paul, and Fred Pearce. 2000. *AAAS Atlas of Population and Environment*. Berkeley: University of California Press.

Hassol, Susan Joy. 2008. Improving how scientists communicate about climate change. *Eos* 89(11):106–107.

Hempel, Carl. 1965. *Aspects of Scientific Explanation, and Other Essays in the Philosophy of Science*. New York: Free Press.

Hoegh-Guldberg, O. 2005. Marine ecosystems and climate change. In *Climate Change and Biodiversity*, edited by T. Lovejoy and L. Hannah, 256–271. New Haven, CT: Yale University Press.

Hoggan, James, and Richard Littlemore. 2009. *Climate Cover-Up: The Crusade to Deny Global Warming*. Vancouver: Greystone Books.

Holland, M. M., and C. M. Bitz. 2003. Polar amplification of climate change in coupled models. *Climate Dynamics* 21:221–232.

Houghton, J. T., Y. Ding, D. J. Griggs, M. Noguer, P. J. van der Linden, X. Dai, K. Maskell, et al., eds. 2001. *Climate Change 2001: The Scientific Basis (Third Assessment Report). Intergovernmental Panel on Climate Change*. Cambridge, UK: Cambridge University Press.

Houghton, J. T., G. J. Jenkins, and J. J. Ephraums, eds. 1990. *Scientific Assessment of Climate Change: Report of Working Group I. Intergovernmental Panel on Climate Change*. Cambridge, UK: Cambridge University Press.

Houghton, J. T., L. G. Meira Filho, B. A. Callander, N. Harris, A. Katteberg, and K. Maskell. 1995. *Climate Change 1995: The Science of Climate Change. Report of Working Group I. Intergovernmental Panel on Climate Change*. Cambridge, UK: Cambridge University Press.

Intergovernmental Panel on Climate Change. 2007a. *Climate Change 2007: Synthesis Report*. Contribution of Working Groups I, II and III to the Fourth Assessment Report of the Intergovernmental Panel on Climate Change [Core Writing Team, Pachauri, R.K. and Reisinger, A. (eds.)]. http://www.ipcc.ch/pdf/assessment-report/ar4/syr/ar4_syr.pdf, accessed September 11, 2013.

Intergovernmental Panel on Climate Change. 2007b. *Contribution of Working Group I to the Fourth Assessment Report of the Intergovernmental Panel on Climate Change, 2007*, edited by Solomon, S., D. Qin, M. Manning, Z. Chen, M. Marquis, K. B. Averyt, M. Tignor, and H. L. Miller. Cambridge, UK: Cambridge University Press and New York: Cambridge University Press.

Intergovernmental Panel on Climate Change. 2012. *Managing the Risks of Extreme Events and Disasters to Advance Climate Change Adaptation* edited by Field, C. B., V. Barros, T. F. Stocker, D. Qin, D. J. Dokken, K. L. Ebi, M. D. Mastrandrea, K. J. Mach, G.-K. Plattner, S. K. Allen, M. Tignor, and P. M. Midgley. Cambridge, UK: Cambridge University Press.

Intergovernmental Panel on Climate Change. 2013a. Organization. http://www.ipcc.ch/organization/organization.shtml, accessed September 15, 2013.

Intergovernmental Panel on Climate Change. 2013b. Working Group I Contribution to the IPCC Fifth Assessment Report Climate Change 2013: The Physical Science Basis Summary for Policymakers. http://www.climatechange2013.org/images/uploads/WGIAR5-SPM_Approved27Sep2013.pdf.

Jacques, Peter J., Riley E. Dunlap, and Mark Freeman. 2008. The organisation of denial: Conservative think tanks and environmental scepticism. *Environmental Politics* 17(3):349–385.

Joint Science Academies. 2008. Joint Science Academies' Statement: Climate Change Adaptation and the Transition to a Low Carbon Society. Academies of Science for the G8+5 countries. http://www.science.org.au/policy/climatechange-g8+5.pdf, accessed September 23, 2013.

Kolbert, Elizabeth. 2006. *Field Notes from a Catastrophe.* New York: Bloomsbury.

Krosnick, Jon A., Allyson I. Holbrook, Laura Lowe, Penny S. Visser. 2006. The origins and consequences of democratic citizens' policy agendas: A study of popular concern about global warming. *Climate Change* 77 (1–2):7–43.

Leiserowitz, A., E. Maibach, C. Roser-Renouf, and J. Hmielowski. 2012. *Global Warming's Six Americas in March 2012 & Nov. 2011.* New Haven, CT: Yale Project on Climate Change Communication. http://environment.yale.edu/climate/files/Six-Americas-March-2012.pdf, accessed July 8, 2013.

Lipton, Peter. 1991. *Inference to the Best Explanation.* Oxford: Routledge.

Lipton, Peter. 2004. *Inference to the Best Explanation.* 2nd ed. Oxford: Routledge.

Lorenzoni, Irene, and Nick F. Pidgeon. 2006. Public views on climate change: European and USA perspectives. *Climatic Change* 77(1–2):73–95.

Manabe, S., and R.J. Stouffer. 1980. Sensitivity of a global climate model to an increase of CO_2 concentration in the atmosphere. *Journal of Geophysical Research* 85(C10):5529–5554.

Manabe, S., and R.J. Stouffer. 1994. Multiple-century response of a coupled ocean-atmosphere model to an increase of atmospheric carbon dioxide. *Journal of Climate* 7(1):5–23.

McCarthy, James J., Osvaldo F. Canziani, Neil A. Leary, David J. Dokken, and Kasey S. White, eds. 2001. *Climate Change 2001: Impacts, Adaptation and Vulnerability. Intergovernmental Panel on Climate Change.* Cambridge, UK: Cambridge University Press.

Metz, Bert, Ogunlade Davidson, Rob Swart, and Jiahua Pan, eds. 2001. *Climate Change 2001: Mitigation. Intergovernmental Panel on Climate Change.* Cambridge, UK: Cambridge University Press.

Mooney, Chris. 2006. *The Republican War on Science.* New York: Basic Books.

Moser, S. C., and L. Dilling. 2004. Making climate hot: Communicating the urgency and challenge of global climate change. *Environment* 10:32–46.

Moser, S. C., and L. Dilling, eds. 2007. *Creating a Climate for Change: Communicating Climate Change and Facilitating Social Change.* Cambridge, UK: Cambridge University Press.

National Academy of Sciences. 2010. *Advancing the Science of Climate Change.* By America's Climate Choices: Panel on Advancing the Science of Climate Change; National Research Council. Washington, DC: National Academies Press.

National Academy of Sciences Committee on the Science of Climate Change. 2001. *Climate Change Science: An Analysis of Some Key Questions.* Washington, DC: National Academy Press.

Olson, Randy. 2009. *Don't Be Such a Scientist: Talking Substance in an Age of Style.* Washington, DC: Island Press.

Oreskes, N. 2004. Beyond the ivory tower: The scientific consensus on climate change. *Science* 306(5702):1686.

Oreskes, Naomi, and Erik M. Conway. 2010. *Merchants of Doubt: How a Handful of Scientists Obscured the Truth on Issues from Tobacco Smoke to Global Warming.* New York: Bloomsbury.

Oreskes, Naomi, Leonard Smith, and David Stainforth. 2010. Adaptation to global warming: Do climate models tell us what we need to know? *Philosophy of Science* 77:1012–1028.

Parmesan, Camille. 2006. Ecological and evolutionary responses to recent climate change. *Annual Review of Ecology, Evolution, and Systematics* 37:637–669.

Planet Ark. 2003. ISS in favor of ExxonMobil global warming proposals. May 19. http://www.planetark.com/dailynewsstory.cfm/newsid/20824/story.htm, accessed July 8, 2013.

Price, Derek de Solla. 1986. *Little Science, Big Science—And Beyond*. New York: Columbia University Press.

Revelle, Roger. 1965. Atmospheric carbon dioxide. In *Restoring the Quality of Our Environment: A Report of the Environmental Pollution Panel, 111–33*. Washington, DC: President's Science Advisory Committee.

Roach, John. 2004. The year global warming got respect. *National Geographic*. December 29. http://news.nationalgeographic.com/news/2004/12/1229_041229_climate_change_consensus.html accessed July 8, 2013.

Rohde, R., R. A. Muller, R. Jacobsen, E. Muller, S. Perlmutter, A. Rosenfeld, J. Wurtele, et al. 2013. A new estimate of the average earth surface land temperature spanning 1753 to 2011. *Geoinformatics & Geostatistics: An Overview* 1:1. doi: 10.4172/gigs.1000101.

Root, T. L., J. T. Price, K. R. Hall, S. H. Schneider, C. Rosenzweig, and J. A. Pounds. 2003. Fingerprints of global warming on wild animals and plants. *Nature* 421:57–60.

The Royal Society. 2005. *A guide to facts and fictions about climate change*. March. http://royalsociety.org/uploadedFiles/Royal_Society_Content/News_and_Issues/Science_Issues/Climate_change/climate_facts_and_fictions.pdf, accessed September 16, 2013.

Somerville, Richard C., and Susan Joy Hassol. 2011. Communicating the science of climate change. *Physics Today* 64:48–53.

Soon, W., S. Baliunas, S. B. Idso, K. Y. Kondratyev, and E. S. Posmentier. 2001. Modeling climatic effects of anthropogenic carbon dioxide emissions: Unknowns and uncertainties. *Climate Research* 18:259–275.

Soon, W., S. Baliunas, S. B. Idso, K. Y. Kondratyev, and E. S. Posmentier. 2002. Modeling climatic effects of anthropogenic carbon dioxide emissions: Unknowns and uncertainties, reply to Risbey. *Climate Research* 22:187–188.

Time. 2006. Americans see a climate problem. March 26.

Union of Concerned Scientists. 2012. *A Climate of Corporate Control: Company Profiles*. http://www.ucsusa.org/scientific_integrity/abuses_of_science/corporate-climate-company-profiles.html.

Van den Hove, Sybille, Marc Le Menestrel, and Henri-Claude de Bettignies. 2003. The oil industry and climate change: Strategies and ethical dilemmas. *Climate Policy* 2:3–18.

Watkin, Daniel J. 2004. Roman Catholic priests' group calls for allowing married clergy members. *New York Times*. April 28, B5.

Watson, Robert T., ed. 2001. *Climate Change 2001: Synthesis Report. Intergovernmental Panel on Climate Change*. Cambridge, UK: Cambridge University Press.

Watson, R. T., Marufu C. Zinyowera, and Richard H. Moss, eds. 1996. *Climate Change 1995: Impacts, Adaptations and Mitigation of Climate Change—Scientific-Technical Analyses. Intergovernmental Panel on Climate Change*. Cambridge, UK: Cambridge University Press.

Weart, Spencer R. 2003. *The Discovery of Global Warming*. Cambridge, MA: Harvard University Press.

Wickham, C., R. Rohde, R. A. Muller, J. Wurtele, J. Curry, D. Groom, R. Jacobsen, et al. 2013. Influence of urban heating on the global temperature land average using rural sites identified from MODIS classifications. *Geoinformatics & Geostatistics: An Overview* 1:2. doi: 10.4172/gigs.1000104.

Wilson, Edward O. 1998. *Consilience: The Unity of Knowledge*. New York: Alfred A. Knopf.

Discussion Questions

1   What pattern has emerged with each successive issue of the IPCC reports?

2   Of the 928 abstracts from scientific research papers during 1993–2003 that Oreskes reviewed, how many refuted the consensus of global climate disruption?

3   What organizations support the statement that "most of the global warming in recent decades can be attributed human activities"? Does their support add credibility to the statement? Why?

4   What factors contribute to much of the American public's perception either that climate disruption does not exist or that human activities do not cause climate disruption? Include the roles of a) mass media, b) scientists, and c) contrarians.

5   Naomi Oreskes describes the process of developing scientific consensus about climate disruption with explanations of the different approaches. Why is it important to understand this before we learn about the details of climate disruption? Describe a) how it helps you individually and b) how it could help the American public. Include how it helps to dispel some misconceptions and how it counters other sources of information.

CHAPTER 4

Climate and Climate Change

Dork Sahagian

We didn't perceive that there was any danger
In being a one-species Earth re-arranger.
Clearing land for our food,
Fossil fuels we thought good,
Turned out as an overall world climate changer.

Weather vs. Climate

Weather is the day-to-day atmospheric conditions, including hourly temperature, precipitation, cloudiness, relative humidity, fog, and the various other atmospheric phenomena we encounter in any given day at any particular point at the Earth's surface. **Climate** is the long-term averaged conditions for a given locality or region. Over periods of decades to centuries, climate controls the nature of vegetation, soils, morphology of the land surface, and the interaction between the land surface, groundwater, surface water, and the atmosphere. Climate in a given locality is controlled most generally by latitude (hot, humid, and rainy at the equator; dry with large diurnal temperature variations at 30° north and south latitudes; cool, cloudy, and rainy at 60° latitude; and very cold and dry at the poles). It is also controlled by elevation, as high mountains are much cooler than low altitudes. Prevailing winds also can determine a region's climate, as hillslopes facing the wind are generally rainier than those facing away from the prevailing winds. (This is because warm, moist air cools as it rises up a slope, thus losing capacity to hold moisture and precipitating out, while heading down the other side, cool dry air gets warmer and thus drier, creating the **orographic effect**). Hawaii is an excellent example of very moist tropical climate on one side of the islands, and very dry desert-like conditions on the other. Yet another factor controlling climate is proximity to an ocean or large lake. Water, due to its great heat capacity, takes much longer to warm up (in summer) or cool off (in winter) than the land surface, so if the ocean is upwind, seasonal temperature variations are greatly moderated. The British Isles are a prime example of the moderating effect of the ocean. In the other extreme, central Siberia and the northern plains of the U.S. experience very large seasonal variations in temperature, with exceedingly hot summers, and very cold winters.

In order to understand climate and climate change, it is necessary to explore the Earth's radiation balance, carbon and other biogeochemical cycles, the hydrologic cycle, all the feedbacks and interactions between Earth's subsystems, and the human impact on the climate system. In this chapter, for both the geologic past (pre-human) and post-industrial, we will investigate climate change on the basis of observations, mechanisms that control climate change, projections of future climate, and expected impacts of climate change on natural and social systems.

Ancient Climate

History of Climate Change

Although climate can be characterized as the long-term atmospheric conditions at a particular location, climate constantly changes at a range of timescales from centuries to tens of millions of years. This has been well documented on the basis of numerous proxies, including glacial ice extent, tree rings, speleothems (stalagmites in caves), ocean and lake sediment cores, isotopes of marine organisms, ice cores, and many other means that geologists use to reveal paleoclimate.

Over the last 600 million years, global climate (averaged over the entire Earth) has varied in temperature from about 12°C to 22°C and back again many times. The cold times are considered an "**icehouse world**," while the warm times are a "**greenhouse world**." Over the last 10,000 years or so, the global average temperature has been about 15°C, and we have been in an interglacial interval of an icehouse world. Throughout Phanerozoic time (last 550 million years), the Earth has oscillated between icehouse and greenhouse conditions about five times, with the most recent greenhouse time being the Cretaceous period, until about 65 million years ago. Since that time, the Earth gradually cooled, culminating in the major glaciations of the Pleistocene Epoch in the last couple of million years. Glaciation waxed and waned throughout the Pleistocene (Figure 4.1), and the most recent major ice sheets that covered northern North America and Eurasia 18,000 years ago essentially melted away by 10,000 years ago. Remnants remain today only in Greenland and Antarctica.

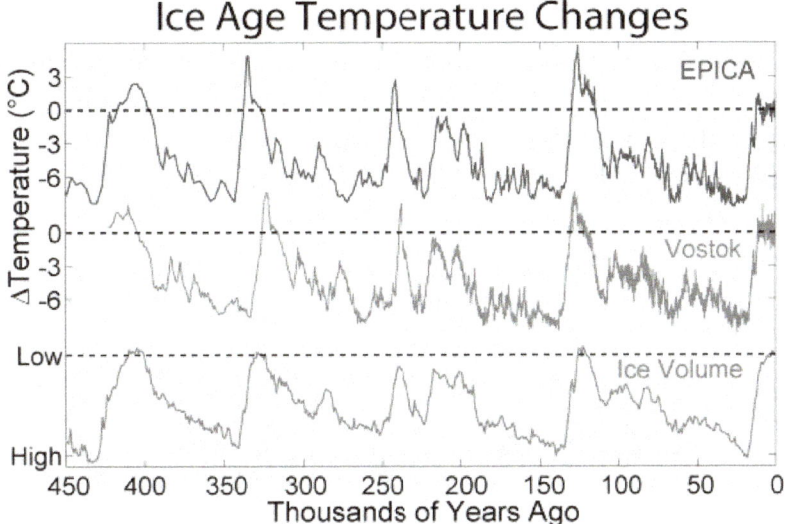

Figure 4.1 The Vostok [and EPICA] ice core[s] drilled from Antarctica preserve[s] a record of climate changes for the latter part of the Pleistocene. Air preserved in bubbles in the ice can be recovered and analyzed for CO2 concentration, isotopic composition, and other characteristics to reconstruct the changing climate over the last 400 thousand years. (This has now been extended back to almost a million years with the same pattern.) Note that the period of oscillation is about 100 thousand years. This is understood as driven by earth's orbital variations [see "Natural Drivers of Climate Change" in text]. (Note also that CO2 concentration oscillated between about 180 ppm and 280 ppm, never rising higher or lower. Apparently some feedbacks within the earth system served to emit CO2 into the atmosphere once it fell to 180 pp m,and draw it back out once it rose to 280 ppm. The details of this are not yet fully understood, and it is an area of active research. However, it is important to note that due to 20th century and current fossil fuel burning, we have raised CO2 concentration to over 400 ppm, far outside the envelope of stability of the last million years or so.)[1]

The Holocene Epoch of the last 12,000 years or so has been a time of remarkable climate stability. (There is a great deal of discussion among scientists about how this time of stable climate may have facilitated agriculture and modern society.) However, even during this stable time, minor climate variations have been recorded. In the last thousand years (Figure 4.3), a Medieval Warm Period occurred in the 12th and 13th centuries A.D., during which the average global temperature was almost a degree warmer than the centuries previous or since, until the 20th century. In the 15th through 18th centuries, the Little Ice Age brought temperatures down to about half a degree below the long-term Holocene average, and Londoners were able to skate on the Thames River, an activity that is quite impossible these days (it doesn't freeze anymore). In the middle of the 19th century, temperatures began to rise and continue to rise at present. The

1   Editor's note: Temperature is determined by using proxy date, i.e., proportions of oxygen isotopes stored in dated layers of ice (see "observing Evidence of Past Climate Change" in text). Note that the ice volume axis is inverted and that this global ice volume is correlated with temperature. The volume is derived from oxygen isotopes in Foraminifera fossils in marine sediment cores, another proxy.

rate of climate warming in the present day is greater than that in the paleo record, and this may be because the temporal resolution of the proxies used to reconstruct past temperatures are insufficient to record such rapid changes, and may be because global warming in the post-industrial era, during which great quantities of anthropogenic greenhouse gases have been released into the atmosphere, is actually more rapid than past natural changes.

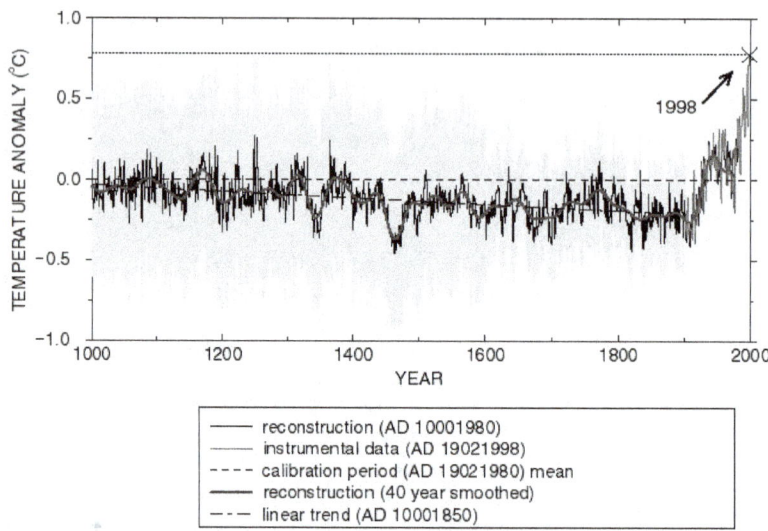

Figure 4.2 Averaged northern hemisphere surface temperature changes since the year 1000. Note that there was slight cooling until about 1900, when temperature began to rise dramatically. It continues to rise, and is projected to rise throughout the 21st century, depending on the rate of fossil fuel burning and associated greenhouse gas emissions. Gray shading indicates range of uncertainty, which decreases for more recent times, as one would expect.

Observing Evidence of Past Climate Change

We were not around to measure climate changes in the ancient past, so we must rely on **proxies** that, by their response to climate change, record these changes and preserve them in the geologic record and other records that scientists can observe and measure. A key observational record comes from ice cores, most notably in Greenland and Antarctica. When snow compacts and recrystallizes to ice, the air trapped between the snowflakes is enclosed in the ice and cannot escape. This air has the composition of the atmosphere at the time the ice formed. By drilling out long ice cores that represent hundreds of thousands of years of ice accumulation, scientists can carefully analyze the air in those bubbles and determine past atmospheric composition. This is how we know how CO_2 and methane changed over time, for example.

Another means for observing evidence of past climates is the analysis of isotopic composition of various organisms, such as marine microorganisms (e.g., foraminifera—"**forams**"). The carbon and oxygen isotopes in their shells reflect the isotopic composition of the ocean surface water from which they made their shells. The oxygen isotopic composition of the water is determined by the amount of glacial ice stored, because high-latitude snow and ice has a much lighter isotopic composition than low-latitude rain has. This is because as water precipitates from vapor to liquid, the heavy isotope (O-18) changes phase into the liquid earlier than the lighter phase (O-16). (The more sluggish heavy molecules want to move more slowly, as in the liquid, rather than energetically bouncing around in the gaseous vapor phase.) This is called **isotopic fractionation**. Because heavy isotopes rain out first, and most evaporation occurs in the tropics, by the time a moist airmass reaches high latitude (or high altitude), it has already rained out the heavy isotopes, and the resulting snow and ice is light, leaving the remaining ocean heavy, so organisms during glacial times have isotopically heavier water to work with. Temperature also affects how organisms incorporate the different isotopes into their bodies, depending on how they do it. Phytoplankton, the base of the marine food chain, use photosynthesis, which strongly fractionates carbon in favor of the light isotope in a way that is sensitive to temperature. In cold water, when there is plenty of dissolved CO_2, phytoplankton can be pickier about preferring the light C isotope, and there is greater fractionation, leaving more heavy C-13 in the water. When zooplankton and larger organisms make their shells from dissolved carbonate, they do not fractionate much, so reflect the composition of the seawater. As such, the degree of **biological fractionation** depends on the temperature of the water and this is reflected in the composition of marine shells. By measuring the isotopic composition of the shells preserved in deep sea cores and rocks, scientists can thus calculate the temperature of the water at the time that the organisms were living.

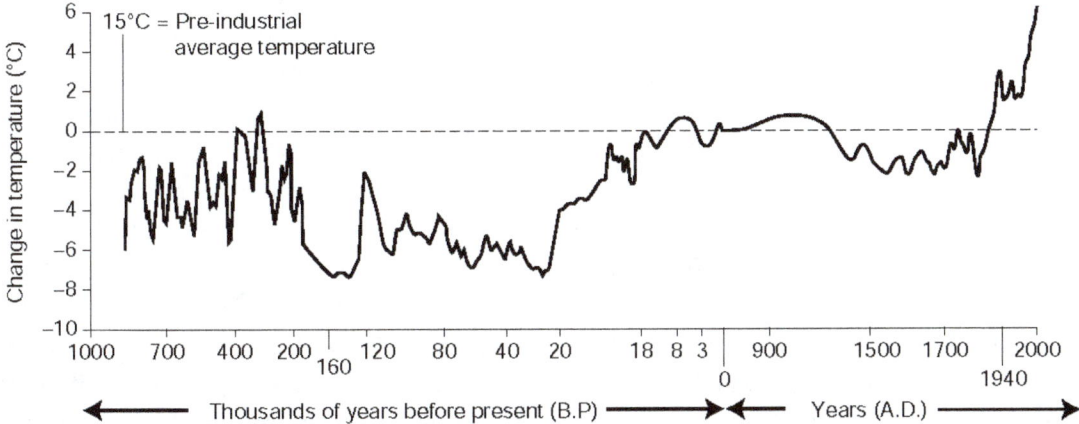

Figure 4.3 History of climate change for last 800 thousand years. Note glacial-intergacial cycles, and modern climate warming. Note also that the scale is non-linear. The ancient past is compressed and the recent is expanded. This has the effect of reducing the apparent rate of recent warming, but it stands out nevertheless.

Yet another proxy for past climate is the nature of **tree rings**. During warm and wet periods, trees grow faster and make thicker rings. Some trees grow very old, and some living Bristlecone pines are up to 5,000 years old, so there is a substantial climate record that can be recovered. By correlating ring patterns from living trees with older, dead trees, the record can be extended even further.

Similar to tree rings but in the marine realm are the growth patterns of **corals**. Annual growth rings can be observed and interpreted regarding the factors that control growth rate in corals (marine chemistry, nutrients, temperature, etc.). In addition, as done for forams and other micro-organisms, corals preserve isotopic ratios that reflect the temperature of the water at the time of growth-ring formation.

Water dripping into caves precipitates calcium carbonate and forms stalagmites, or **speleo-thems**. Because the water comes from rain and flows through the land surface and rocks above the cave, the isotopic microstructure of the speleothems reflects the oxygen composition of the rain water as well as the carbon fractionation processes of the terrestrial ecosystem overlying the cave, with temporal resolution that can be as fine as weeks. It has even been discovered that some speleothems can record the passing of individual hurricanes in the tropics, and used as a measure of hurricane frequency in the past.

Natural Drivers of Climate Change

At the longest timescales of tens of millions of years, climate can be affected by the motions of the Earth's tectonic plates, moving land areas to more polar or more equatorial positions. In more polar positions, glaciation and the associated ice-albedo feedback can lead to global cooling, while equatorial land and polar oceans lead to more moderate general climates. On shorter timescales, however (less than a million years), too short for tectonic motions to matter, there are other mechanisms for climate change. The most notable of these is the variations in the Earth's orbit around the Sun. These were discovered in the 1920s by Milutin **Milankovitch**, who was able to relate orbital variations to geologic evidence of the timing of Pleistocene glaciations. There are three timescales of orbital variability (Figure 4.4). The first and longest, with a periodicity of roughly 100,000 years, is the shape of the orbit, which is always elliptical, but with varying **eccentricity**—sometimes more circular, sometimes less. The second is the tilt of the axis relative to the plane of the orbit (known as **obliquity**) with a period of about 41,000 years, varying between 21.5 and 24.5 degrees (current obliquity is 23.5 degrees, the latitude of the tropics of Cancer and Capricorn). The third, **precession**, relates to the interaction between the former two, because the rotational axis wobbles so that sometimes northern summer occurs when the Earth is closest to the Sun (in its eccentric orbit), and sometimes, as at present, north-ern summer occurs when it is farthest from the Sun, thus reducing seasonality (recall that most of the Earth's land surface is in the northern hemisphere). The period for precession is about

23,000 years. All three Milankovitch cycles affect global climate, but sometimes one frequency more strongly controls glacial–interglacial cycles, and sometimes another frequency dominates. It is not entirely clear why this should be so.

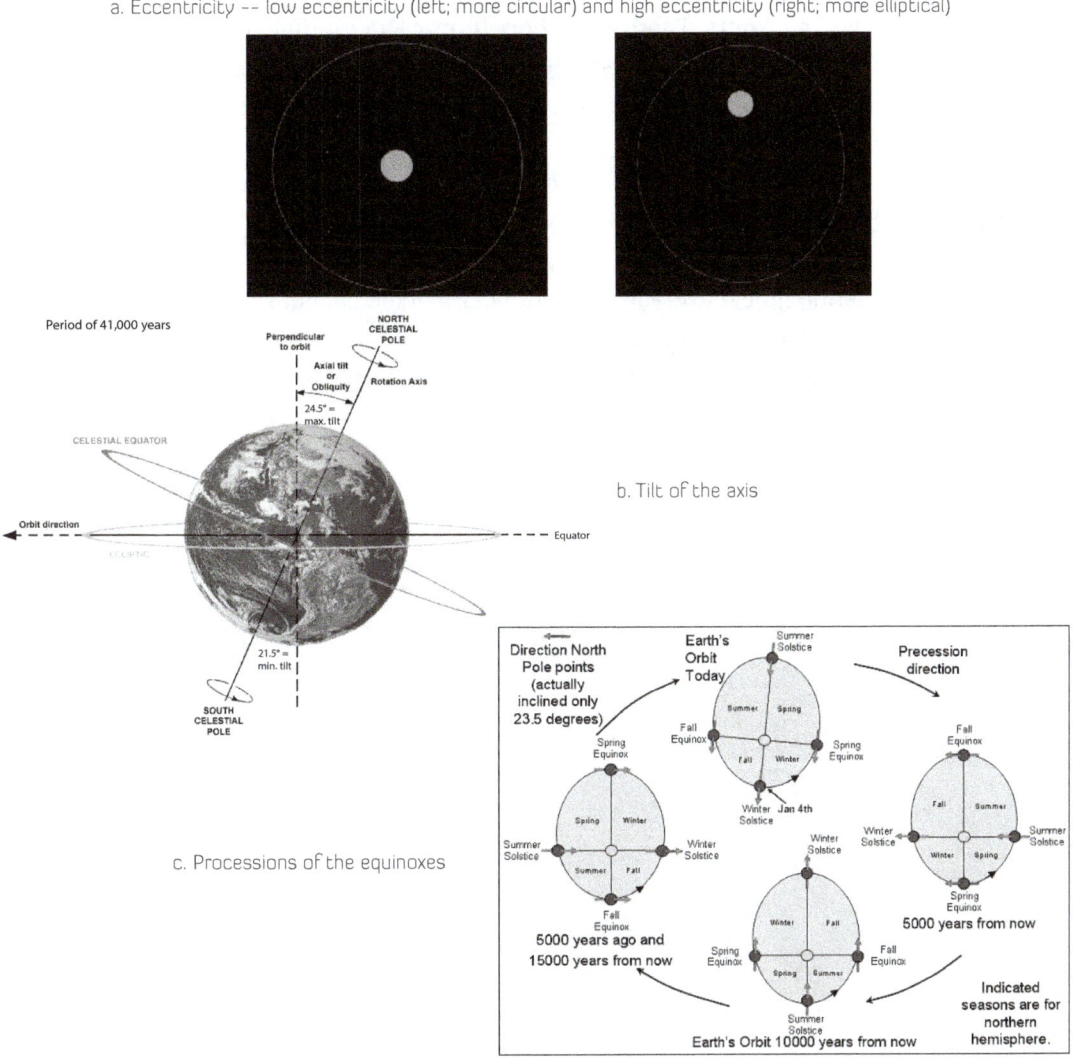

Figure 4.4   Milankovitch cycles. Long-term climate is affected by earth's periodically varying orbital parameters. Eccentricity is the "roundness" of the orbit around the sun (the sun remains at one focus of the elliptical orbit). At present, the orbit is fairly round. The tilt of the earth's rotational axis also changes, defining the tropics of Cancer and Capricorn as well as the Arctic and Antarctic circles. At present, the tilt is a moderate 23.5 degrees. Finally, the interaction between the eccentricity and tilt changes (precesses), so that the earth is closer or farther from the sun during northern hemisphere summer. At present, we are farther from the sun in northern hemisphere summer, thus moderating climate slightly. The last several glacial-interglacial cycles were driven at the eccentricity period of about 100,000 years.

On shorter timescales, **solar cycles** can influence climate, as they control solar output, reflected in the number of sunspots. There have been times, such as the **Medieval Warm Period**, when there were an abundance of **sunspots** (solar storms) that emitted more solar energy, and times such as the Maunder Minimum that corresponds with the **Little Ice Age**, when there were few sunspots. There is a short-term cyclicity that results in variations in sunspot numbers, which oscillate between 200 and just a few in a cycle that has been about 11 years in period ever since it has been recorded, but it appears to be shortening to about 10 years recently.

Other natural variations occur within the Earth's climate system, and **El Niño Southern Oscillation (ENSO)** is a prime example. Every 3–7 years, the warm pool of equatorial Pacific surface water moves east and west, altering the global atmospheric circulation patterns, and while not significantly altering global average temperatures, marked changes in regional climates are observed throughout the world. Other oscillations include the North Atlantic Oscillation, and the Pacific Decadal Oscillation, each of which affects regional and global climates in a cyclic fashion with a period of several years to decades.

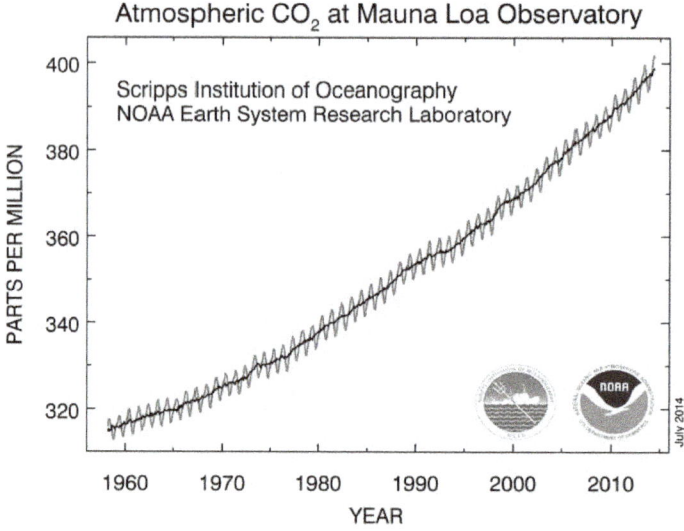

Figure 4.5 The record of atmospheric CO_2 concentration on top of Mauna Loa, on Hawaii in the middle of the biggest ocean on earth, far from most industrial or other sources of anthropogenic CO_2. Note that the rate of increase has been increasing since 1958, when recording began. The thick line is the annually averaged CO_2 concentration. The thin line varies annually as the northern hemisphere plants bloom in the spring and decay in the Fall and winter. This can be likened to the "breathing" of the northern hemisphere, where most deciduous land plants live.

Modern Climate

Observations of Recent Climate Change

The issue of rapid recent and projected climate change has risen from an obscure topic for specialized academic scientists to explore to a household word that has come to the fore of the political arena. Many consider it the most serious environmental problem that humanity has ever faced. Although humans have been conducting activities, such as ecosystem destruction for agriculture, hunting, and biomass burning, that alter the Earth system for millennia, the impacts

Twenty Warmest and Coolest Years on Record (°C Anomaly from 1901–2000 Mean)

The tables below are sorted by temperature anomaly, or the amount by which the average temperature of each year exceeds the average for the entire century. It is remarkable that the warmest years (left) are recent, while the coolest years (right) were a century ago.

Table 1 a) 20 hottest years. b) 20 coldest years.

a.			b.		
2005	0.6183		1909	−0.4179	
2010	0.6171		1908	−0.4144	
1998	0.5984		1911	−0.4131	
2003	0.5832		1904	−0.3999	
2002	0.5762		1910	−0.3999	
2006	0.5623		1907	−0.3702	
2009	0.5591		1912	−0.3544	
2007	0.5509		1893	−0.3459	
2004	0.5441		1903	−0.3453	
2001	0.5188		1913	−0.3340	
2011	0.5124		1917	−0.3333	
2008	0.4842		1894	−0.3095	
1997	0.4799		1892	−0.3087	
1999	0.4210		1890	−0.3082	
1995	0.4097		1916	−0.2949	
2000	0.3899		1905	−0.2759	
1990	0.3879		1891	−0.2654	
1991	0.3380		1898	−0.2585	
1988	0.3028		1887	−0.2516	
1987	0.2991		1895	−0.2503	

on global systems were relatively minor until the industrial revolution, when huge quantities of fossil fuels began to be mined and burned. As a result, various gases were emitted into the atmosphere, including oxides of sulfur and nitrogen, and most notably, carbon dioxide. While the sulfur and nitrogen led to acid rain that was subsequently alleviated by scrubbers and other end-of-pipe technologies, there has been to date no satisfactory means for capturing and permanently storing carbon dioxide. Consequently, the atmospheric concentration of CO_2 has risen from the pre-industrial level of 280 ppm, to a 2014 level of over 400 ppm, and it continues to increase at an accelerating rate, currently about 2 ppm per year (Figure 4.5). Because CO_2 is a greenhouse gas, this has triggered the modern era of global warming and associated climate changes and impacts throughout the world. While there is a long list of other greenhouse gases that have even greater global warming potential per molecule, the sheer quantity of CO_2 emitted to the atmosphere by fossil fuel burning has caused it to be the dominant driver of recent rapid climate change.

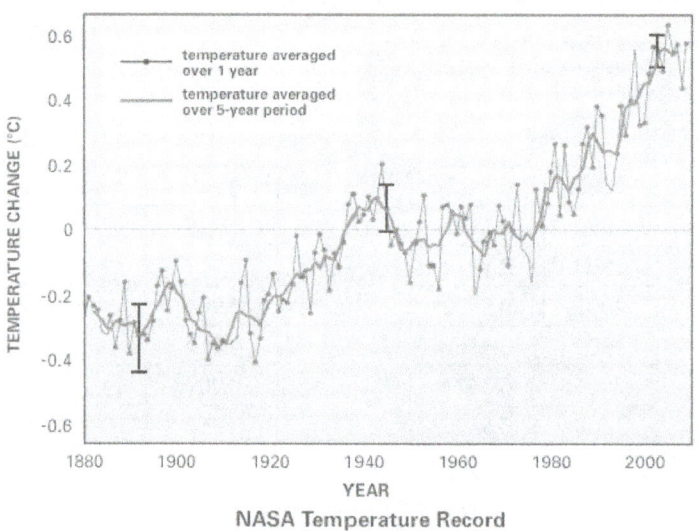

Figure 4.6   Observed global averaged temperature as a departure from the 1951–1980 average.

A remarkable observation is that of average global temperatures of the 20th century. If you sort the 20 warmest years of the century, you see that they are all quite recent (Table 4.1). If you list the coolest 20 years, they are all early in the century. In fact, 13 of the warmest 14 years ever recorded have been in the 21st century. Of the 10 warmest years ever, 9 of them have been in the 21st century, and the tenth was 1998. The post-industrial history of global temperature shows dramatic warming (Figure 4.6). This observation settles any question about whether or not there is any global warming.

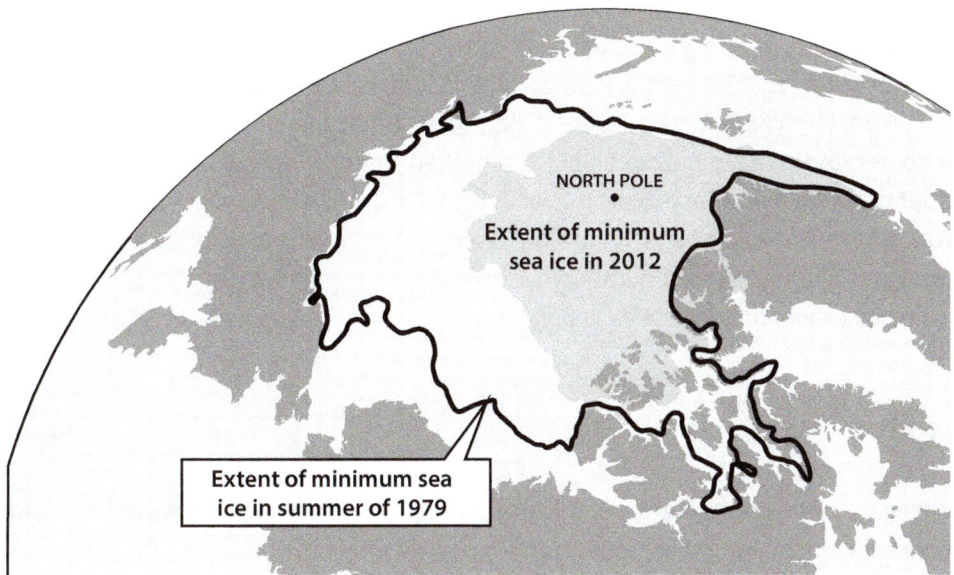

Figure 4.7 Arctic sea ice has dramatically declined in the last several years. This triggers a strong positive feedback from ice albedo (near 1) and exposed ocean water (albedo near 0). The more that ice melts to expose water, the more solar energy is absorbed, melting more ice, exposing and warming more water, etc. The rate of Arctic sea ice loss was severely underestimated by the scientific community as reflected in the Fourth Report of the Intergovernmental Panel on Climate Change (IPCC).

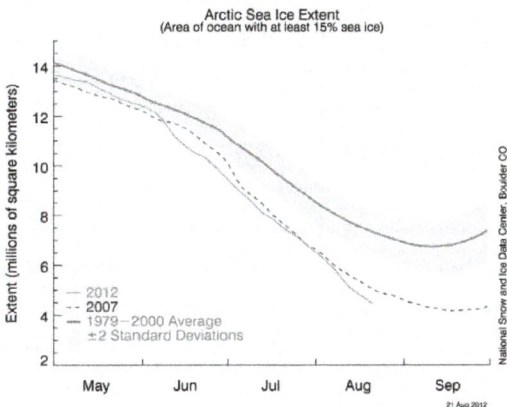

Figure 4.8

In response to observed climate changes, there has been an outcry from a number of international organizations as well as the scientific community to reduce and shortly thereafter curtail fossil fuel burning. This may, at least at first, involve necessary reductions in energy consumption, until sufficient non-carbon-based energy sources are developed.

There has been a great deal of confusion in the public realm, particularly in the U.S., about the causes of observed 20th century warming (and 21st), and the role of human activities in the climate system. This confusion has arisen from the fact that there are a great many factors that control observable climate, including, but not limited to, sun spots, ocean currents, cycles such as ENSO, other regional effects, volcanic activity, land use changes, marine biological activity, and many others. These complexities are the realm of Earth System Science, a very complicated business, indeed. It has taken generations of scientific investigation just to begin to understand what is involved, but some headway is being made.

Arctic Ice

An alarming observation in recent years is the deterioration of sea ice in the Arctic Ocean. While it has been long understood that global warming would affect the polar regions more profoundly than the tropics, no prediction even came close to predicting the rate of Arctic ice disintegration that has been observed in the last several years. The famous ice minimum of late September, 2007, was observed just a month before the monumental volume on climate change was published by the Intergovernmental Panel on Climate Change (IPCC), such that the volume was out of date before it was even released. It did not predict the huge decrease in Arctic ice that was observed a month before it was released, after years of research and production. The minimum ice in 2011 came close, but 2012 surpassed all, setting a new record for minimum ice ([Figures 4.7–4.9]). In 2012, there was very little multi-year ice remaining. Almost the entire Arctic will now be covered by first-year ice, meaning new, thin, recently formed ice, rather than thick leftover ice that has persisted for many years. Even the maximum extent of Arctic ice, observed each February, is declining. The Artic is now navigable for a significant part of the summer (by regular boats—not icebreakers). This has never been observed in recorded human history.

The global impact of declining Arctic ice stems from the strong **ice-albedo effect**, causing incident sunlight to be absorbed by an ocean with albedo near zero rather than reflected by snow and ice with albedo near one. This positive feedback has already set in, and we may see further declines in summer ice and an extended ice-free season in the Arctic, with thinner and thinner winter ice in future years.

Sea Level

Sea level has varied by hundreds of meters over geologic time, and is rising rapidly today. Observations of 20th century sea level rise have been made from tide gauges in the world's harbors. Although originally installed to measure the diurnal and semi-diurnal times for shipping

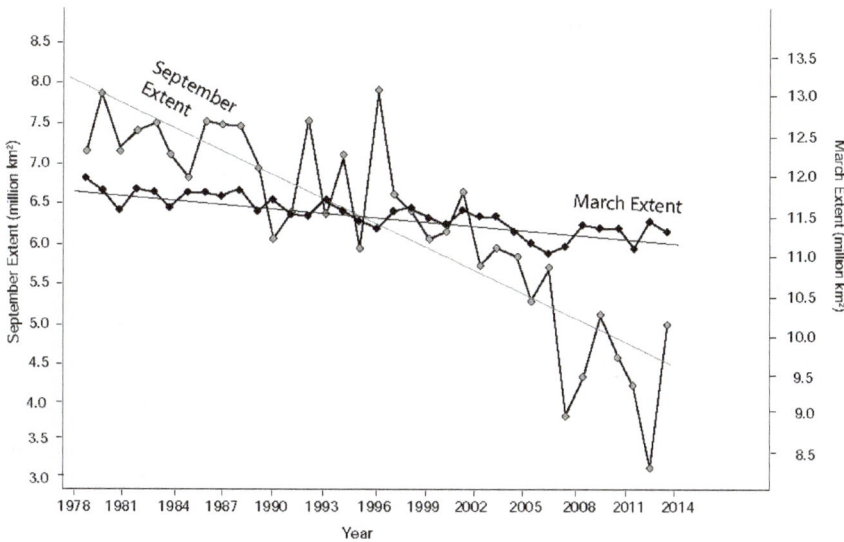

Figure 4.9 Historical variation of summer sea ice in the Arctic Ocean. The minimum of Arctic sea ice cover each year is observed in September, and a dramatic minimum occurred in 2007. In 2008, some thought that a partial recovery may have occurred, but in 2010, there was another dramatic decline, and in 2011, the areal extent was almost as small as in 2007. The all-time minimum so far was in September, 2012, with a slight increase in 2013. It is noteworthy that record lows have never been set two years in a row (after each record low, there is a slight rebound). This may be caused by the albedo of open water vs. ice. Water's very low albedo absorbs almost all energy (like a "black body"), but it also emits energy much more readily than high-albedo ice. So in the dark winter of a minimum ice season, all that open water radiating to space loses more energy than would be lost by water insulated conductively, and especially by radiation, by ice. Note that the minimum in September is followed by an uptick in March ice cover the following calendar year. This would lead to greater ice cover the next September as well, so you don't see two record low Septembers in a row. However the long-term trend continues to decline. This has led to projections of an ice-free Arctic during the summers in as little as 10–15 years. Due to the severity of the ice-albedo positive feedback, such projections are difficult to quantify, and have been found to underestimate rates and magnitudes when applied to other Earth system parameters. This feedback may also be causing the average age of Arctic ice to be declining, with more first-year ice, and less multi-year ice each year. On that basis, we would expect the more rapid decline in summer ice than in winter ice, as can be discerned from these figures from the National Snow and Ice Data Center. Note that until 2007, the time of maximum ice extent was in February, but since 2007, the (declining) annual maximum has been occurring later, in March.

purposes in the ports, over the long term, it was noticed that the average level of the tides was rising—sea level rise. After correcting some locations for tectonic motions, including delayed vertical rebound of the Earth's crust due to deglaciation, a 20th century rate of sea level rise was observed to average about 1.5–1.75 mm/yr. In the second decade of the 21st century, we now have a long enough satellite altimetry record to observe that the rate of sea level rise has increased to about 3 mm/yr.

Drivers of Recent (and Near Future) Climate Change

Greenhouse Gas Emissions

The primary driver of recent (20th–21st century) climate change is anthropogenic emissions of greenhouse gases. These emissions have been dominated by CO_2 resulting from fossil fuel burning, but also include methane, and numerous other trace gases that contribute to climate warming due to trapping of outgoing infrared radiation.

The greenhouse effect has already been described in Chapter 12, and for the purposes of this chapter, it will suffice to say that when greenhouse gases are added to the atmosphere, a temperature increase cannot be avoided by any means. The correlation between observed greenhouse gas emissions and observed climate change is no accident. Despite delays in the response of the climate system to greenhouse gas forcing, most of the observed 20th century warming trend has been found to be due to anthropogenic emissions (Figure 4.10). Some climate contrarians maintain that greenhouse gas emissions have not altered global climate. Clearly this is wrong, but more importantly, the delayed response of the climate system to the 20th century emissions is not so severe that we have yet to observe associated warming. If warming had not yet started due to delayed response, all impact of greenhouse gas emissions would occur in the future, above and beyond what we have observed so far, leading to far more severe climate impacts than actually expected.

In 1996, the Second IPCC Report stated that "The balance of evidence suggests that there is a discernible human influence on global climate." In 2001, the Third Report stated that "There is a new and stronger evidence that most of the warming observed over the last 50 years is attributable to human activity." In 2007, the most definitive report to date, the Fourth Assessment Report stated that "Most of the observed increase in globally averaged temperatures since the mid-20th century is very likely due to the observed increase in anthropogenic greenhouse gas concentrations." Most recently, in 2013, the Fifth Report stated that, "It is *extremely likely* that human influence has been the dominant cause of the observed warming since the mid-20th century." While the scientific conclusions have remained essentially unchanged since the early 1990s, the mounting scientific evidence clearly demonstrates the magnitude of the human influence on climate changes through greenhouse gas emissions, as exacerbated by land use changes.

CO_2-Water Feedback

A major area of confusion among the American public is the role of water vapor as a greenhouse gas. Anthropogenic greenhouse gases are being emitted to the atmosphere as a result of human activity—mostly agricultural and industrial. The greenhouse warming caused by these gases serves to trigger an increase in the most important greenhouse gas—water vapor. Water vapor is responsible for most of the 33°C of the total greenhouse effect on the Earth. Because warmer air can hold more water vapor (think of hot, humid weather relative to cold, dry days), there is a strong positive feedback that forms a vicious cycle of warming, once the atmosphere is warmed

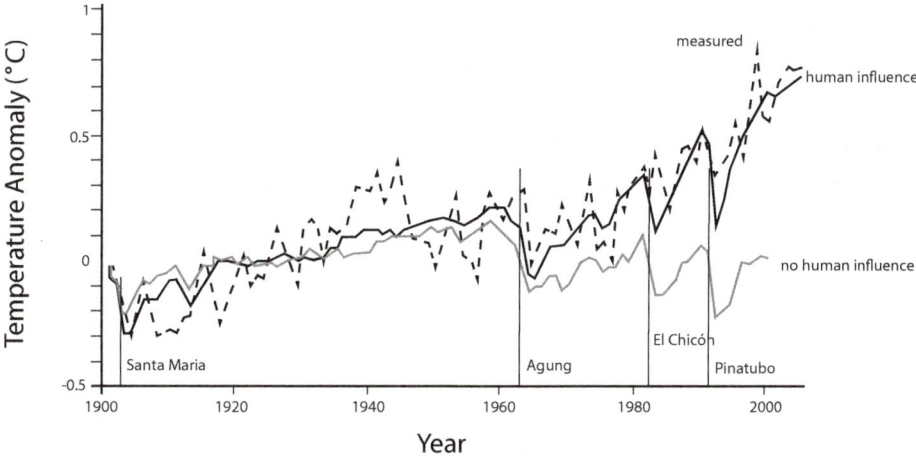

Figure 4.10   Reconstructions from numerical models compared to instrumental observations of average global temperatures since the industrial revolution. Observed climate change can be attributed to a combination of natural and anthropogenic forcing of the system. When compared to observed temperature history, models that account for only natural climate variability miss the observed rapid recent warming, but when anthropogenic influences are included, the observed data are fit most closely. It is clear from these curves that the anthropogenic influence is much greater than the natural variation of climate, which shows no appreciable warming over the course of the 20th century.

a little by other means. During the glacial–interglacial climate oscillations, this external driver was orbital variations that warmed the atmosphere slightly, allowing the water vapor effect to take over and cause much more warming. Since industrial times, the trigger has been the anthropogenic greenhouse gases described here (mostly CO_2). By causing a very slight amount of warming directly, these gases enable the atmosphere to hold more water vapor, which in turn causes a great deal more warming, which, in turn, allows the atmosphere to hold yet more water vapor, leading to additional warming, etc. Some "climate contrarians" misunderstand the role of water and CO_2 and contend that because CO_2 is a mere trace gas, anthropogenic emissions have no significant effect on climate. When climate contrarians claim that CO_2 has no effect on climate because during glacial–interglacial cycles warming preceded CO_2 increase, they are missing the point that the initial warming was caused by orbital variations and then greatly amplified by the water vapor greenhouse effect. The warming of the atmosphere then warmed the ocean, which, when warmed, could not hold as much CO_2 in solution, so CO_2 exsolved from the water and went into the atmosphere, and this is recorded in ice cores. The additional direct greenhouse effect from the CO_2 itself would have been small relative to water vapor. However, the CO_2 NOW being added to the atmosphere (from fossil fuel sources) is sufficient to trigger the water vapor effect and lead to marked global warming and related climatic changes. Water has a very short residence time in the atmosphere, as it literally rains out frequently. CO_2, on the other hand, has a very long residence time, on the order of millennia, so its effects cannot be removed quickly, and we have only begun to feel the warming triggered by CO_2.

Some who have not studied the details of planetary energy balance may also fall victim to the **CO_2 saturation fallacy**. Some climate contrarians claim that CO_2 already absorbs all of the outgoing energy within its absorption spectrum, and thus promote the fallacy. In fact, absorption of infrared by CO_2 can only happen for discrete wavelengths, according to the energy levels of the vibrational modes of the molecule. Thus there is not enough CO_2 in the atmosphere to absorb all the outgoing infrared in the broad range of absorption. Near the ground the high temperature and pressure of the lower troposphere causes intermolecular collisions that serve to broaden absorption bands, but only slightly, still leaving large gaps for plenty of infrared wavelengths to slip through. Further, not only is CO_2 not saturated in the earth's atmosphere (as it actually is on Venus), but even if it were saturated for infrared absorption as a whole, it is only the top portion of the atmosphere (that is even further from saturated) that emits radiation to space. Lower in the atmosphere, CO_2 (and water) molecules absorb and then radiate infrared radiation in all directions (up AND down). So when you add CO_2 to the atmosphere, you get a thicker "blanket" for absorbing and radiating, with warming at all levels, while the level from which radiation escapes to space rises higher. Moreover, although water vapor is the most important greenhouse gas near the ground, the upper atmosphere is dry so water does not play a role there. However, CO_2 is well mixed throughout the entire atmosphere, and thus dominates the radiative balance at the top of the atmosphere where it controls planetary radiation out to space.

The problem is that the atmosphere should not be imagined as a single piece of matter, like the glass of a greenhouse for which the effect is named. As such, the "greenhouse effect" is not really an accurate view of what happens. When the ground (and water) surface of the earth emits infrared radiation, much of it is absorbed by the greenhouse gases in the air near the ground. These gases then re-radiate infrared in all directions, including up to overlying air and back down to the ground. This sets up a chain reaction of absorption and re-radiation that eventually reaches the top of the atmosphere. By adding CO_2 to the atmosphere, a deepening amount of it engages in absorption and re-radiation, thus effectively thickening the "blanket" of greenhouse gases on the planet and moves the location that the atmosphere radiates out to space to a higher altitude, leading to greater warming at lower levels in order to maintain radiative equilibrium.

20th Century Aerosols

In the mid-20th century, the industrialized nations of North America and Europe burned a great deal of high-sulfur coal, thus emitting large amounts of sulfur dioxide into the atmosphere. This formed aerosols of SO_2 that served to reflect incoming solar energy back into space before reaching the ground. Ultimately, this served to cool the lower troposphere (where thermometers reside on the ground) in these large regions, in addition to causing severe acid rain and surface waters in particularly susceptible areas. In response to the acid rain problem, technologies were developed to scrub out the sulfur from coal-fired power plant smokestacks, greatly reducing sulfur emissions. This removed the cooling effect of the aerosols as well. Many climate contrarians who wish to obfuscate the scientific discussion of climate change have cited the mid-20th century cooling during a time of rapid fossil fuel burning as "proof" that climate change is not caused by CO_2 emissions from fossil fuel burning. However, they do not understand (or admit) the local cooling effect of aerosols, which was temporary in any case. Now that we do not have that cooling effect, climate is warming everywhere, including on the ground in industrialized nations.

An additional source of aerosols is specific volcanic eruptions. In a notable 1991 eruption of Pinatubo, tons of sulfur (as SO_2) erupted into the atmosphere, and this had a temporary cooling effect on global climate. For a couple of years, temperatures were measurably cooler than in previous years. However, like industrially produced aerosols, they quickly rained out, and climate was restored to the normal of the time. Although sulfur aerosols have a temporary cooling effect on climate, no one is seriously advocating adding more sulfate aerosols to the atmosphere in any attempt to mitigate global warming. The acid rain produced is a deadly impact of sulfate aerosols, as experienced by the many dead lakes in the Adirondack Mountains of New York and many other places. Further, the cooling effect of aerosols is local, temporary, and functions only during the daylight hours, while the global warming effect of CO_2 operates 24/7, as the Earth emits infrared all the time. Consequently, aerosols cannot be used to effectively mitigate the climate changes triggered by anthropogenic CO_2 emissions.

Carbon and Climate

Because anthropogenic CO_2 plays such an important role in the climate system, the carbon cycle has become a critical area of research for climate scientists. As discussed in Chapter 6, the natural carbon cycle includes short-and long-term sources and sinks, such as annual northern hemisphere spring bloom, weathering of silicate rocks, the solubility and biological marine pumps, subduction of carbonate sediments, and CO_2 emission from volcanic eruptions. These processes, while cycling vast quantities of carbon, maintain a steady state, such that the CO_2 that resides in the atmosphere is kept at reasonably constant concentrations, varying between glacial and interglacial times between 180 and 280 ppm. With the advent of fossil fuel burning, we have removed enough geologically stored carbon from deep reservoirs and thus released enough CO_2 to raise atmospheric concentration to 400 ppm, and rising at an accelerating rate currently at 2 ppm per year. This has suddenly brought atmospheric chemistry out of equilibrium with global climate, ocean temperature and circulation, biome distribution (although this is also perturbed by direct land use), and ice cover. The Earth may not have ever seen such a severe level of disequilibrium—certainly not as long as humans have existed as a species, and the process of re-equilibration will involve some fundamental shifts in the operation of the Earth system, including (in addition to general warming), storm intensification, glacial melting, sea level rise, precipitation changes (with wet places getting wetter and dry places drier in general), and perhaps most significantly, **ocean acidification**, as discussed in Chapter 9. Such rapid CO_2 increase with such cold ocean water makes it possible for the ocean to dissolve greater concentrations of CO_2 than in previous times of high atmospheric CO_2 because the ocean has not had time to warm up yet in response to global warming, and thus can hold more carbon, and thus become more acidic than perhaps ever before.

Impacts of Climate Change

Temperature

Temperature is the primary parameter one thinks of in discussions of climate change (Figure 4.11), but it is by no means the only change, and, in fact is of less real concern than various other parameters that are driven by changing thermal structure and energy balance of the planet. Indeed, temperature variations between glacial and interglacial cycles were only a few degrees. It changes more than that every day and night, and much, much more than that seasonally. People would not notice an average temperature change of the expected four or five degrees over the course of the 21st century, when it changes much more than that every day and night. It is the impacts of other parameters that are very sensitive to temperature that are of greater direct concern.

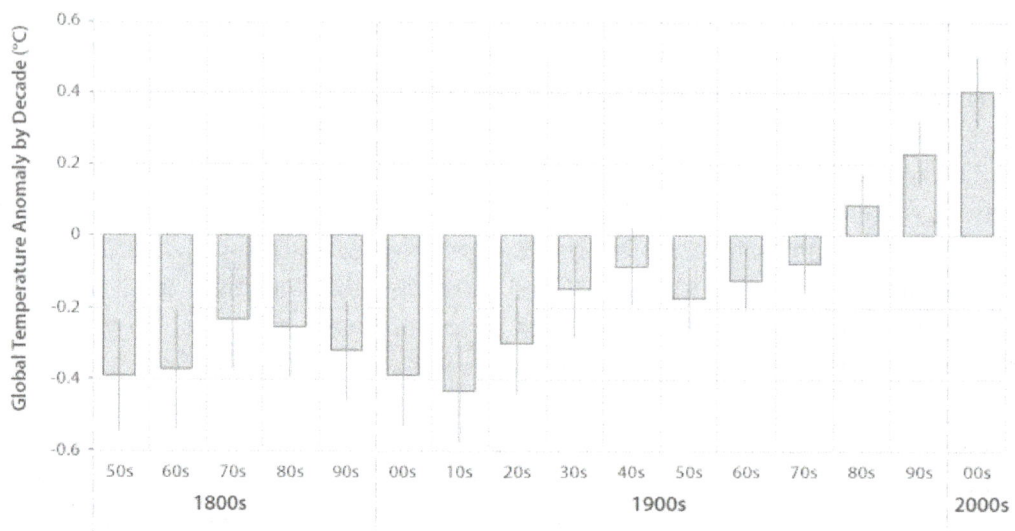

Figure 4.11 Post-industrial temperature record, with baseline in 1980. Note rapid warming in the late 20th century. The warming trend that started in the early 1900's was interrupted by the emission of sulfur aerosols in mid-century that led to acid rain and were subsequently halted by using scrubbers in coal-burning power plants, after which warming quickly caught up to its 20th century trend. Vertical lines represent range of uncertainty.

Precipitation Patterns

Climate models indicate that the changes in precipitation patterns will depend on the extent of warming caused by different potential scenarios of anthropogenic greenhouse gas emissions. In a business-as-usual scenario (A2 or RCP 8.5 of IPCC), severe changes should be expected, with relatively dry places generally getting drier and relatively wet places getting wetter. The implications of this are more droughts where agriculture may already be water-limited, and more floods where additional water is not needed. In a scenario in which emissions are rapidly curtailed (B1 or RCP 2.6 of IPCC), this effect would be greatly reduced (but still present). A discussion of what the world would look like under the various IPCC scenarios may become moot, however. The A2 and B1 scenarios were defined almost 20 years ago, and sufficient time has passed already to observe that we are actually going along the path of the extreme emissions case of the business-as-usual A2 case. In 2013, about 36 gigatons of carbon dioxide were emitted by human activities (equivalent to about 9 gigatons of just carbon—do the stoichiometry), near the A2 scenario. As precipitation patterns change in the 21st century, adaptation measures will need to be taken to ensure that food production can be maintained, and that areas subject to additional flooding and droughts bolster infrastructure and community organization to reduce vulnerability.

Storms

Large cyclonic storms such as **hurricanes** are driven by energy entrained by the evaporation of warm sea surface water. (This is why they happen most frequently in late summer and fall, when ocean temperatures are highest.) The warmer the water, the more energy there is to fuel the storm. Global warming involves heating of the ocean as well as the atmosphere (in fact the ocean water is far more important due to its heat capacity), so we should expect more serious storms as a result. Global warming does not make hurricanes. They occurred in pre-industrial times. However, ocean temperature plays a strong role in determining the nature of hurricanes. There has been considerable discussion in the scientific community regarding the impact of climate change on large storm systems. It has now been reasonably well established that warming will not spawn more storms, but that once the cyclonic flow is generated in the tropical North Atlantic, for instance, with increasingly warm water to fuel them, the storms will grow more powerful and maintain their strength longer as they travel farther north. Coastal cities will be braced for more severe impacts, as seen in New Orleans in 2005 (Katrina; Rita) and New York in 2012 (Sandy).

Sea Level Rise

One of the most troubling impacts associated with climate change is its effect on sea level. With the rate of sea level rise increasing from between 1.5 and 1.75 mm/yr in the 20th century, to more than 3 mm/yr in the 21st century, there is concern for the security of many of the world's coastal cities. Problems will be felt first during storms, such as Hurricane Sandy that struck New York and New Jersey. There is a double whammy involved here. As storms become more intense (but not more frequent), the increased storm surge and wave heights will act on top of an already higher base level due to sea level rise. When the storm surge is over a couple of meters, every additional centimeter counts in terms of flooding large areas, inundating and filling tunnels, subways, and other infrastructure, and the rate and extent of transport of coastal sediments such as the sands of beaches.

Ocean Acidification

Anthropogenic CO_2 emissions do more than contribute to the greenhouse effect and cause global warming. As explained in Chapter 9, in order to maintain chemical equilibrium between the ocean and the atmosphere, some of the excess atmospheric CO_2 enters solution in the ocean. This leads to [lower] pH and makes it more difficult for marine organisms to make shells out of carbonate dissolved in the water. If this ocean sink (the biological pump) weakens, more emitted CO_2 will remain in the atmosphere, contributing further to greenhouse warming. As ocean water warms in response, CO_2 becomes less soluble, slowing (and potentially reversing) the solubility pump as well. There has not been a great deal of concern to date regarding ocean acidification, but will become a topic of increasing interest in the coming years.

The Intergovernmental Panel on Climate Change (IPCC)

In response to the potential threats climate change poses not only to human systems, but also to the functioning of all aspects of the global ecosystem, the governments of the world formed the Intergovernmental Panel on Climate Change (IPCC). This body serves to bring together scientists from all over the world to address the difficult science of understanding the climate system and predicting climate change for the coming century and beyond. IPCC was established by the World Meteorological Organization (WMO) and the United Nations Environment Programme (UNEP) to assess scientific, technical, and socio-economic information relevant for the understanding of climate change, and its potential impacts and options for adaptation and mitigation. It is open to all members of the UN and of WMO. It bases its assessment on peer-reviewed and published scientific/technical literature. The goal of IPCC is to use scientific literature to evaluate the extent and understanding of climate changes, as well as the potential to adapt to or counteract climate changes.

The IPCC organizes and reviews the published findings of the world's leading scientists, convenes workshops, produces detailed reports on various topics, and has provided four major Assessment Reports starting in 1990. Although the scientific community's understanding of the climate system progressed a great deal in the intervening two decades, the projections and predictions have changed remarkably little since the early years, indicating that even then, enough was known about the climate system for governments to begin to take action to mitigate future climate change.

IPCC reports were published in 1990, 1995, 2001, 2007, and 2013. While earlier reports were less detailed and based on fewer years of scientific research, they all resulted in the same basic

BASIC QUESTIONS OF CLIMATE CHANGE

Despite the complexities of global climate change, there is one aspect of Earth's climate that is relatively simple to understand, and that is the balance of energy between the Sun and the Earth's surface. Sunlight comes through a fairly transparent atmosphere and heats the surface; that heat is radiated back to space as infrared radiation. However, the atmosphere is not quite as transparent to infrared, so absorbs this energy, warming the atmosphere. This process is the greenhouse effect. If we make the atmosphere more or less transparent to infrared radiation by adding or subtracting certain gases, we can cool or warm it directly. These effects are very simple physics, and are well understood, explained in introductory textbooks such as this one. Further, CO_2 is a trace gas that warms the atmosphere, thus enabling it to hold more water vapor, the dominant greenhouse gas. So CO_2 acts as a trigger to load more water vapor, which warms more, thus holding more water, warming more. On top of that, because warm water cannot hold in solution as much CO_2 as cold water can (try opening a warm seltzer!), warming atmosphere warms the ocean, which then could begin to exsolve some of its CO_2 into the atmosphere, warming further. This runaway feedback between CO_2, temperature, and water vapor is the primary concern regarding CO_2 and other greenhouse gas emissions. The details of these processes are explained in the comprehensive report of the Intergovernmental Panel on Climate Change (IPCC, 2007), which summarizes and double checks the scholarly publications of the international scientific community in area of climate change. While there may be a great deal of discussion within the scientific community about the details of the interactions between the various components of the Earth system, the basic physics of radiation and absorption has been settled for over a century. In sum, if CO_2 or other greenhouse gases are added to the atmosphere, it must get warmer.

Regarding anthropogenic climate change there are these four basic questions:

1. **Is climate changing?** Yes. There are undeniable instrumental data that temperatures are rising, precipitation patterns are changing, and ocean and atmospheric circulation systems are changing throughout the 20th and 21st centuries.

2. **Do people have anything to do with it?** Yes. Greenhouse gas emissions (primarily CO_2 from fossil fuel burning) have to warm the atmosphere—it is what they do. The consensus of model results shows that the global climate is sufficiently sensitive to historic anthropogenic CO_2 emissions to have already warmed by the amount measured over the last 150 years.

3. **Is climate change bad?** Yes. While this is a more normative question to be considered by philosophers and the general public rather than by scientists, history has shown that any change in the environment of stable civilizations is disruptive to those civilizations. Alterations in areas in which crops can be grown, changes in phenology (when plants bloom or flower and leaves fall, and when insects emerge, etc.), shifting storm tracks, and rising sea level may have devastating economic, social, and political consequences to modern societies.

4. **Can we do anything about it?** Yes. Because much of the warming caused by past emissions has already occurred, cessation of emissions can stabilize climate in the 21st century. Until they are overwhelmed, natural carbon sinks in the ocean and terrestrial ecosystems can continue to absorb previously emitted carbon and return global climate to the stable state in which civilization evolved over the last 10,000 years.

What we learn from the past is that nearly every major climate change in Earth's history has been accompanied by changes in greenhouse gases, with warming associated with more CO_2 and cooling associated with less. In the geologic past, before humans existed, climate and atmospheric CO_2 concentrations varied together, with CO_2 change not always predating climate change. This was due to the runaway feedbacks between temperature, CO_2 in the atmosphere and ocean, and water vapor in the atmosphere. However, now that we have devised a way to inject CO_2 directly into the atmosphere (fossil fuel burning), CO_2 is preceding climate warming, which is already responding to the additional greenhouse gases.

conclusions regarding observation, attribution, and projection of climate changes. However, most predictions of the previous reports have proven to be underestimates of both the rates and magnitudes of climate change and its impacts on ice cover, sea level, ecosystems, and other aspects of the Earth system. The Fourth Assessment Report, published in 2007, included greater detail than any previous report, and gained the notice of more people than any report that preceded it. In addition, it served as the basis for the book and documentary film *An Inconvenient Truth*, which brought the science of climate change into more homes and discussions than had all the scientists in the world for the preceding 20 years. For this, and for the continuing exhaustive work of the IPCC and all the scientists involved, Al Gore and the entire IPCC were awarded the Nobel Peace Prize in 2007. (Al Gore also received an Oscar for the film in the same year.) In the past, some Nobel Prizes were awarded for work already completed and problems solved. In this case, however, the scientific community is still in the process of identifying the extent of the problems associated with climate change, and active research continues in the areas of understanding drivers, responses, and interconnections within the climate system, impacts of climate change on terrestrial and marine ecosystems, and implications for alterations in climate and ecosystems for human society. The Fifth Assessment Report of the IPCC was released in Fall, 2013. The basic conclusions are the same as previous IPCC reports, but uncertainties have been reduced, and some 21st century predictions of rates and amounts of change for temperature, regional variability, and sea level rise have been increased as our models and observations have improved in recent years. While the basic scientific conclusions may not differ greatly from the previous reports, the added evidence and reductions in uncertainty of predictions of changes and responses of the various parts of the Earth system may help steer the governments of the world who have organized the IPCC to take the most effective action possible to both mitigate climate change and adapt to its impacts. The general strategy is to mitigate what you can, and adapt to the rest. In the simple words of the UN Scientific Export Group Report on Climate Change and Sustainable Development, and later the Union of Concerned Scientists, "Avoid the unmanageable so that you can manage the unavoidable."

The following sections are provided as background for understanding the science of climate change. The most comprehensive and up-to-date treatment of climate change is the 2013 report of the IPCC. It was written in several parts, the most relevant of which is the report of Working Group I, "The Scientific Basis." IPCC also published a "Technical Summary," and a "Summary for Policymakers," both of which provide a more digestible synopsis of the findings of the international scientific community than the full report. The entire report is several thousand pages long; even the first part on "The Scientific Basis" weighs in at over 2000 pages, providing plenty of reading for those who wish to explore every detail of the science of climate change. Rather than reproducing this landmark work within this introductory textbook, because it is free for all to access online, the links are provided here for students to obtain the relevant parts of the reports directly from IPCC, starting with the "Summary for Policymakers" and the "Technical Summary" of Working Group I.

The first reading in this section is the "Summary for Policymakers" found at

http://www.ipcc.ch/report/ar5/

The summary provides a short introduction to the science of climate change as well as a few of the key results that are policy-relevant.

With the publication of the 2013 AR5, the 2007 report became out of date, although the basic science remains unchanged. For example, in the 2007 report, it was indicated that 11 of the 12 years between 1995 and 2006 ranked among the 12 warmest years since 1850, when instrumental records began to be reliably kept. In 2012, this was updated to "20 of the last 25 years (1987–2011) were the warmest years in the 20th century." And now, in 2014, we see that of the warmest 14 years ever recorded, 13 of them were in the 21st century (and the 14th was 1998)! The warming trend continues unabated despite well-discussed but completely unimplemented measures to reduce greenhouse gas emissions and land use changes.

The next part of the IPCC to be read is the "Technical Summary," which provides the scientific basis for the conclusions summarized in the "Summary for Policymakers."

Those interested in a fully comprehensive treatment of the problem can obtain the full report of Working group 1 (and 2 and 3) at the IPCC website. The "Synthesis Report" should be available soon.

Key Terms

Biological fractionation: Isotopic fractionation by organisms during growth. Photosynthetic pathways (both C3 and C4) strongly prefer the light isotope of carbon (both in land plants and marine phytoplankton). Phytoplankton thus deplete the surface ocean in C-12 (relative to C-13) "enriching" the water, so that when planktonic organisms make their shells from the remaining inorganic C, they are enriched (isotopically heavy) in C-13 relative to the deeper ocean.

Black Body: An object that absorbs and emits every wavelength of electromagnetic radiation perfectly (0 albedo). Black bodies heat in the presence of incident radiation (and cool in its absence) more quickly than objects that do not absorb and emit radiation as efficiently. The ocean is close to a black body, while floating ice is close to its opposite, a white body.

CO_2 saturation fallacy: The impression (by some who know just a little about science) that there is enough CO_2 in the atmosphere to absorb ALL outgoing infrared radiation in the entire range of the CO_2 absorption spectrum, so adding more CO_2 wouldn't make a difference. However, a molecule absorbs radiation of a specific set of wavelengths, and even in the high pressure of the lower troposphere, where the specific absorption bands are spread out due to collisions and interactions between molecules, there are wide gaps for radiation of not-quite-the-right-wavelengths to escape. So even the lower atmosphere is not saturated in the sense of absorbing all the infrared. Furthermore, the atmosphere is not like a sheet of greenhouse glass—rather, it absorbs and re-radiates infrared in all directions at every level of the atmosphere. So it is the concentration of CO_2 at the top of the atmosphere (where there is no water, which is limited to troposphere) that matters, and if you add CO_2, it will increase, thickening the "insulating" blanket of greenhouse gases surrounding the planet, necessarily causing warming at all levels of the atmosphere.

Eccentricity: The non-roundness of an ellipse (or in this case, elliptical orbit), as measured in terms of the ratio of major and minor semiaxes as $(1-(b^2/a^2))^{1/2}$—a circle has eccentricity 0, and a very elongated ellipse has eccentricity approaching 1. Earth's orbital eccentricity waxes and wanes with a period of about 100,000 years.

El Niño Southern Oscillation (ENSO): Oceanographic and atmospheric conditions caused by a shift of a warm pool of equatorial Pacific surface water from the western side of the Pacific to the eastern side, closer to South America. This occurs every 3–7 years and suppresses upwelling of nutrient-rich deep waters off the coast of South America. Without nutrients, phytoplankton do not bloom as profusely, thus affecting the higher trophic levels, reducing the utility of the fishery off South America.

Forams: (short for foraminifera). A single-class phylum of the kingdom Protista that primarily lives at the ocean bottom or surface, and makes its shells from calcium carbonate, whose isotopic composition reflects the temperature as well as isotopic composition of the ocean at the time that organism lived.

Greenhouse world: Warm conditions globally, during which there was little polar ice, resulting in a warm deep ocean.

Hurricanes: Cyclonic (anticyclonic in the northern hemisphere) storms triggered by tropical atmospheric instabilities in the absence of wind shear, and fueled by the latent heat of evaporating seawater.

Ice-albedo effect: The strong feedback between snow/ice cover and solar heating. The sun reflects away from high-albedo snow and ice, thus not heating it, so snow makes it colder, keeping things cold. If there is a reduction in snow/ice cover, this heats the low-albedo land surface or ocean water much more, thus heating more, reducing the snow/ice cover, and heating more. This positive feedback creates an instability that can lead to rapid loss or gain of ice. Currently we are losing Arctic ice very quickly.

Icehouse world: Globally cold conditions, during which extensive continental glaciers (ice sheets) waxed and waned (glacial-interglacial cycles within icehouse times). We are in an icehouse world now, in an interglacial period, resulting in a cold deep ocean.

Isotopic fractionation: The preferential inclusion of a light or heavy isotope of an element during phase change of a substance. The heavier isotope "prefers" liquid over gas, and solid over liquid. Therefore, when water evaporates, for example, heavy isotopes of O and H are left behind in the ocean.

Little ice age (17th to 19th centuries, A.D.): A time of few sunspots, attributed to a cooling climate.

Medieval warm period (11th to 14th centuries, A.D.): A time of many sunspots that are attributed to a warmer than usual climate on Earth.

Obliquity: The tilt of the Earth's rotational axis relative to the plane of its orbit around the sun. It increases and decreases with period of about 41,000 years.

Ocean acidification: The reduction of pH of ocean water by solution of excess atmospheric CO_2 through the solubility pump. Historical pH was 8.2, but if it falls below 8.0, marine organisms will have difficulty making calcium carbonate shells, thus threatening the biological pump, as well as the marine ecosystem. With a cold ocean, as present, and with sudden high atmospheric CO_2 concentrations (as we are tending toward due to fossil fuel burning), more CO_2 can be dissolved in the ocean than would naturally be possible, leading to concern regarding the marine ecosystem.

Precession: The relation between eccentricity and obliquity, pertaining to whether the Earth is nearer the sun during the northern hemisphere's summer or winter. Presently, it is closer to the sun in northern hemisphere winter, thus slightly reducing seasonality.

Proxies: Records of past environmental conditions preserved in sediments, rocks, trees, ice, and any other remnants from the time of interest. The records may be chemical, biological, isotopic, and physical, and could be preserved in any number of ways. Scientists use these records to infer past conditions, because they understand that the processes that create the records depends on the past environmental conditions that created them.

Solar cycles: Periodic variability in the number of sunspots on the sun, with a period about about 11 years. Sunspots produce bursts of energy that temporarily provide greater insolation to the Earth (warming).

Tree rings: Annual growth rings in trees. They are normally wider (faster growth) during warm and especially wet times.

Credits

Discussion Questions

1 Why is it illogical to invoke earth's orbital changes as a cause of modern climate disruption?

2 Why are deposits of microscopic Foraminifera so helpful in reconstructing ancient temperatures of Earth?

3 What is one of the most reliable sources of ancient CO_2 records?

4 Compared to a 30-year average over recent decades, what has happened to the extent of Arctic Sea ice during the past six years?

5 What is the CO_2-water feedback, and why is it important?

CHAPTER 5

Climate Change and Sea-Level Rise

Gary Griggs

Every nation on Earth with a coastline should be worried about sea-level rise. Throughout virtually the entire history of civilization, the past eight thousand years, more or less, depending on how we define civilization, the level of the ocean, and therefore the location of the shoreline, did not change much. While the tides did go in and out every day and storms and hurricanes temporarily elevated sea levels regionally, the overall level of the ocean stayed pretty much the same.

This started to change in a significant way about 150 years ago as humans began to use fossil fuels in increasingly larger quantities. Burning coal, oil, and natural gas, as well as wood, peat, and animal dung, produces carbon dioxide. With 85 percent of our global energy being produced from burning fossils fuels, the generation of carbon dioxide (CO_2) is still increasing. In 2014 our global population emitted about 40 billion tons of CO_2, twice as much as in 1980 and ten times as much as in 1930. Forty-four percent of this came from just two countries, China and the United States. Of that yearly production, China is now the global leader, with 28 percent of the total; the United States is second with 16 percent; and the European Union generates just 10 percent. India and the Russian Federation each produce 6 percent; Japan, 4 percent; and all the other nations, the remaining 30 percent (Figure 5.1). From 2005 to 2014 about 44 percent of that carbon dioxide accumulated in the atmosphere, 26 percent in the ocean, and 30 percent on land, primarily in vegetation. While we are also emitting methane and nitrous oxide, CO_2 makes up 76 percent of greenhouse gases.

Global temperature is directly responsible for the Earth's climate, which in turn determines how much of the planet's water is locked up in ice sheets and glaciers and how much is contained in the oceans as salt water. Global temperature also affects the temperature of the ocean and how

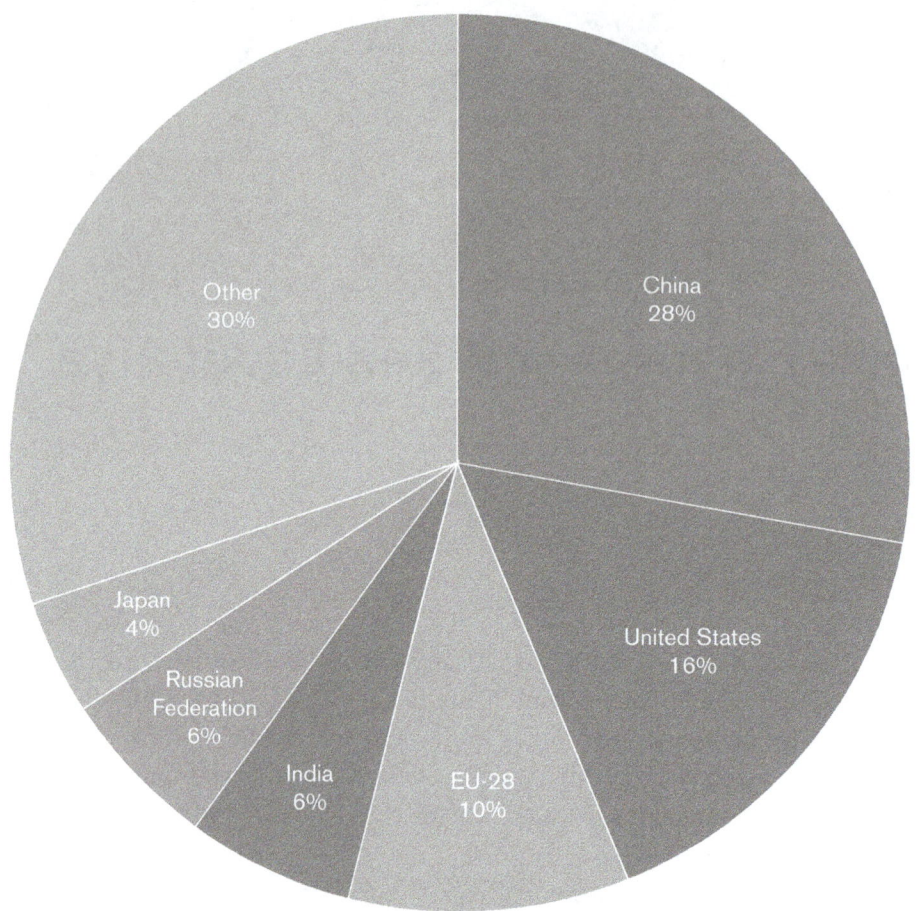

Figure 5.1   Percentage of global emissions of carbon dioxide by country (2015).

much volume it takes up. As the planet warms, as it has been doing since the last ice age ended about 18,000 years ago, ice has been melting and the oceans have been gradually warming, with both processes conspiring to slowly raise sea level. At the end of the last ice age about 10 million cubic miles of ocean water was tied up as ice sheets and glaciers, which lowered sea level nearly 400 feet. The subsequent postglacial warming raised sea level in fits and starts to its present elevation.

Sea-Level Change: Driving Forces

The climate and temperature of Earth are influenced primarily by the amount of heat it gets from the sun, which is directly related to Earth's distance from the sun at any period in time. The Serbian engineer Milutin Milankovitch recognized nearly 150 years ago that the distance between the Earth and the sun changes over time because of irregularities in the Earth's orbit, as

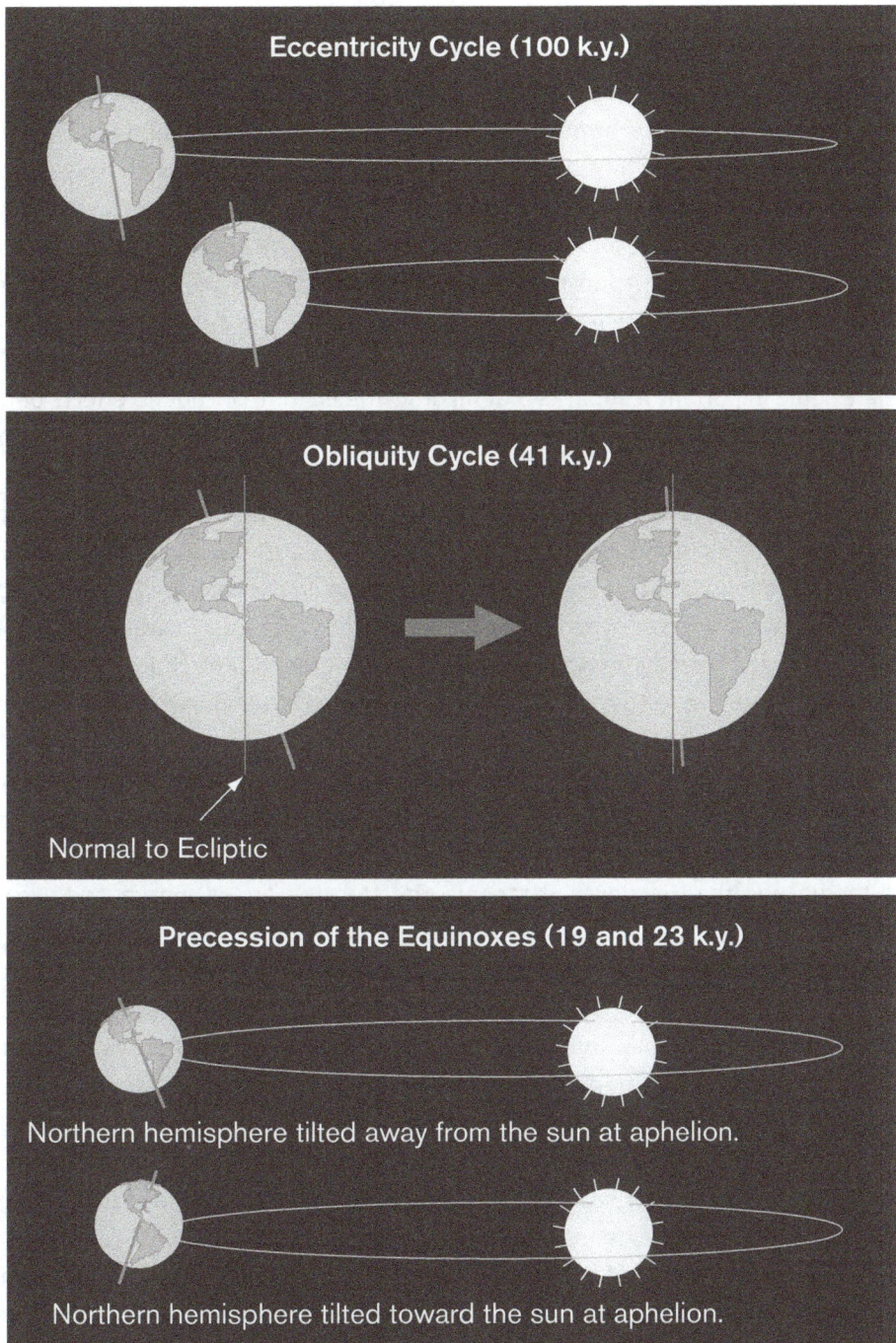

Figure 5.2   The three irregularities in the Earth's rotation and in its orbit around the sun, known as the Milankovitch Cycles (KY = thousands of years).

well as the tilt and wobble of the Earth on its axis (Figure 5.2). As the axis of the Earth's rotation tilts a few degrees, or its orbit moves it farther away from the sun, we receive less solar energy, and as a result the Earth and its atmosphere and oceans cool slightly. When the Earth is closer to the sun, global temperatures increase. There are well-understood cycles in these orbital oscillations that span tens of thousands of years and are associated with well-documented warming and cooling intervals that have profoundly affected the Earth's surface, its ice cover and the ocean volume, and therefore the location of the shoreline. These orbital differences in distance are not huge but have been sufficiently large to lower or raise the Earth's temperature enough to help either initiate or terminate an ice age.

There are some important feedbacks that magnify the heating or cooling effects of these orbital variations. When the Earth begins to enter a warmer, or interglacial, phase, three important processes are affected. First, as more shelf ice or floating Arctic ice melts, more ocean surface is exposed to sunlight. The ocean surface absorbs heat, in contrast to ice, which reflects heat. Second, permafrost at high latitudes starts to thaw as the Earth begins to warm, and in doing so it releases carbon dioxide and methane from the decaying and formerly frozen vegetation it contains. And third, a warmer ocean cannot hold as much CO_2 as cold water, so as the ocean warms CO_2 is released to the atmosphere, adding to the greenhouse effect and causing more warming. Each of these processes can also move in a reverse direction, when the Earth starts to cool as a result of its relationship to the sun.

All other things being equal, and they rarely are, the warmer it gets, the more ice will melt and the higher sea level will rise. There is only so much ice on the surface of the Earth, however, so there is a limit to the extent sea level can rise from climate change. The mountain glaciers of the Earth, those in the Himalayas, Patagonia, the Alps, the Andes, Alaska, and even Glacier National Park, while impressive in their own right, are relatively small in area and total volume. Virtually all of them have retreated significantly over the past 150 years. If these glaciers continue to retreat and were to completely melt, they would add only about 2 feet to the total level of the oceans. While relatively small compared to the two other large storehouses of ice, this could be very significant if you live within 2 feet of sea level, and lots of people around the planet do.

Greenland is an entirely different story, however. If its ice cover were to completely melt or calve off into the ocean, sea level would rise about 24 feet. Greenland is melting at an increasing rate (Figure 5.3), and 24 feet of sea-level rise would inundate many densely populated areas around the world that are situated virtually at sea level. A number of large cities are already experiencing seawater in some of their streets now, in a process often called tidal or nuisance flooding. Along the Atlantic seaboard, these include Boston, New Haven, Atlantic City, Baltimore, Ocean City, Norfolk, Charleston, Savannah, and Miami, to name a few (Figure 5.4). Future sea-level rise will increase the frequency of this wet inconvenience in very predictable ways (Figure 5.5). Communities bordering San Francisco Bay also experience tidal flooding today, as do some Southern California beach communities. Globally, over 150 million people live in parts of cities within 3 vertical feet of high tide, including London, Alexandria, Kolkata, Mumbai, Dhaka, Yangon,

Figure 5.3   Meltwater river and moulin (a deep circular pit where meltwater enters a glacier) on the Greenland Ice Sheet, 2010. (Photo: Adam Scott © 2010, www.adamscottimages.com)

Figure 5.4   Nuisance flooding during high tides is becoming increasingly more common in many cities along the Atlantic Coast of the United States. At Annapolis, Maryland, in December 2012 wind, rain, and high tides combined to create significant flooding. (Photo: Amy McGovern licensed under CC BY 2.0 via Flickr).

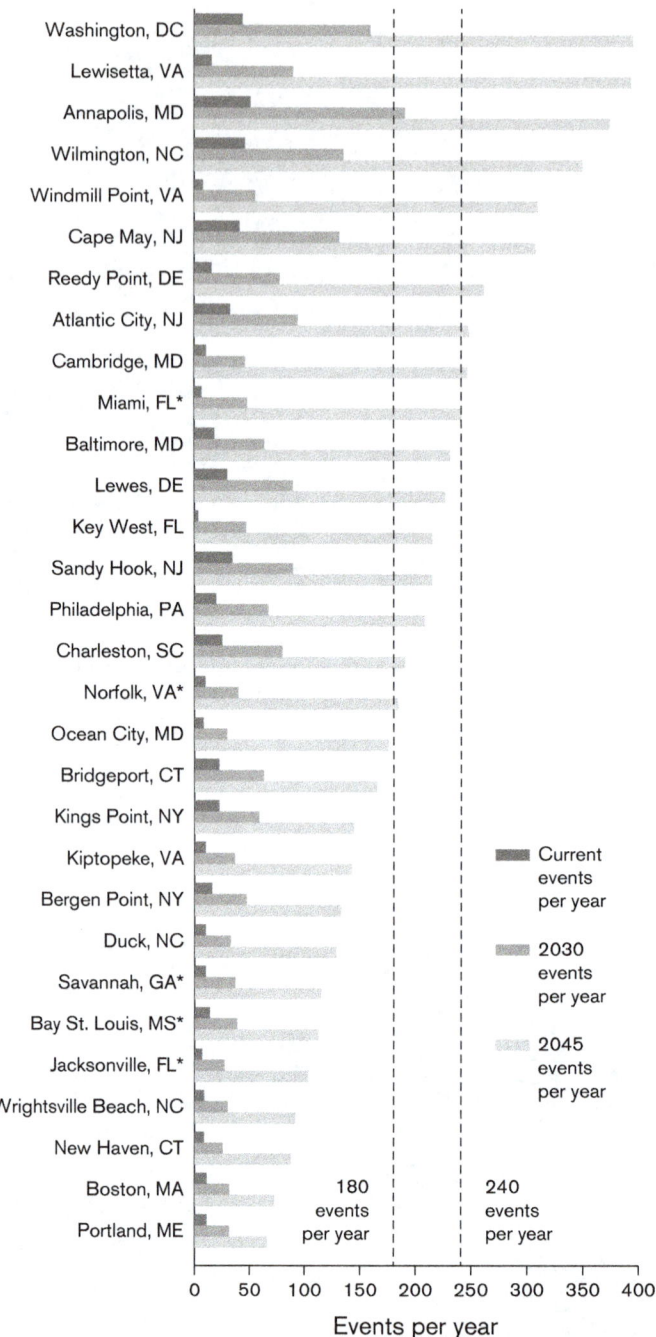

Figure 5.5 Frequency of tidal flooding in cities along the Atlantic Coast of the United States in 2012 and projections for 2030 and 2045. (Courtesy of Union of Concerned Scientists, *Encroaching Tides,* www.ucsusa.org).

Bangkok, Jakarta, Ho Chi Minh City, Hai Phong, Shanghai, Guangzhou, and Tokyo, to name a handful. This tidal flooding is discussed later in this chapter.

Antarctica has by far the largest volume of ice, about 70 percent of all the freshwater on the planet, and enough to raise sea level globally by about 200 feet were it all to melt. The ice shelves surrounding parts of Antarctica are melting at rates that are historically unprecedented, but the largest volume of ice in Antarctica lies in the interior of the continent, locked in by mountains, so it is not all likely to melt anytime soon. On the other hand, scientists who study Antarctica have serious concerns about the potential for one or more of the very large glaciers to accelerate and break off in the decades ahead, which could lead to a very rapid rise in sea level of several feet or more. The floating ice shelves are presently acting as buttresses or corks to keep the glaciers from advancing quickly. However, with continued ocean warming and sea-level rise, the concern is that these ice shelves will break up, allowing the very large glaciers to advance and calve off into the sea.

While global cycles of heating and cooling, or interglacial and glacial periods, have been going on for several millions of years, these natural fluctuations have been altered in a major way during the past 150 years by the increasing concentrations of greenhouse gases in the atmosphere from human activities, primarily fossil fuel combustion (~91 percent of contribution) and land use changes such as the burning of tropical rain forests and other vegetation (~9 percent).

The Earth's atmosphere has always contained natural greenhouse gases, primarily carbon dioxide, methane, and nitrous oxide, and these have made the difference between a barely habitable planet and one where life as we know it can thrive. Without this natural greenhouse effect, the average temperature on the planet would have been about 0°F (−18°C) rather than the 60°F (15°C) we enjoy. The greenhouse effect is not a new idea; it goes back over a century to the work of the Swedish chemist and Nobel Prize winner Svante Arrhenius. But human activity has been adding to the natural greenhouse content of the atmosphere for well over a century, and these anthropogenic increases have been well documented.

The longest records of the changes in the greenhouse gas content of the atmosphere have been recovered in recent years from the ice of Antarctica. As snow falls and accumulates on the ice fields of the southern continent, air bubbles are trapped between the snowflakes and are then frozen into the ice. By drilling down through the ice, now as deep as 11,000 feet below the surface, continuous ice cores have been recovered that extend back over 800,000 years. The chemistry of these preserved air bubbles has been analyzed and reveals that atmospheric CO_2 concentrations fluctuated between about 180 and 300 parts per million (ppm) over this long period of prehuman history (Figure 5.6) but have increased significantly over the past century.

Almost halfway around the world, on the flank of Moana Loa on the island of Hawai'i, daily measurements of the carbon dioxide concentration in the atmosphere document a 27 percent increase since the late 1950s. By 2016 the atmospheric CO_2 concentration in Hawai'i had reached 405 ppm (Figure 5.7). This represents a 43 percent increase over the natural levels that persisted for hundreds of thousands of years that are preserved in Antarctic ice. There is no longer a scientific question that human activities have enhanced the natural greenhouse effect, leading to a warming of the Earth of nearly 1.5°F, and that this trend continues to increase. The warming of the Earth is causing the oceans to heat up and more ice to melt, which combine to raise sea level.

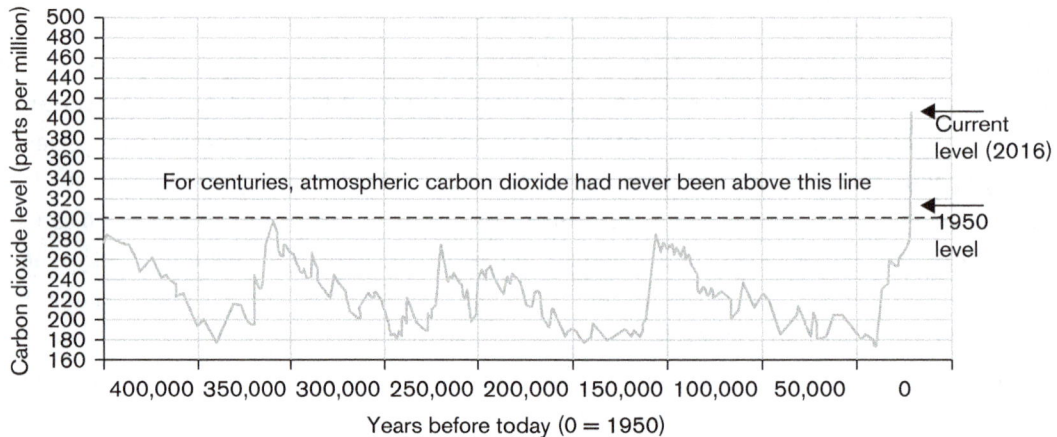

Figure 5.6 The carbon dioxide content of the atmosphere is now significantly higher than it has been at any time during the past 400,000 years as recorded in Antarctic ice cores. (Image: NASA).

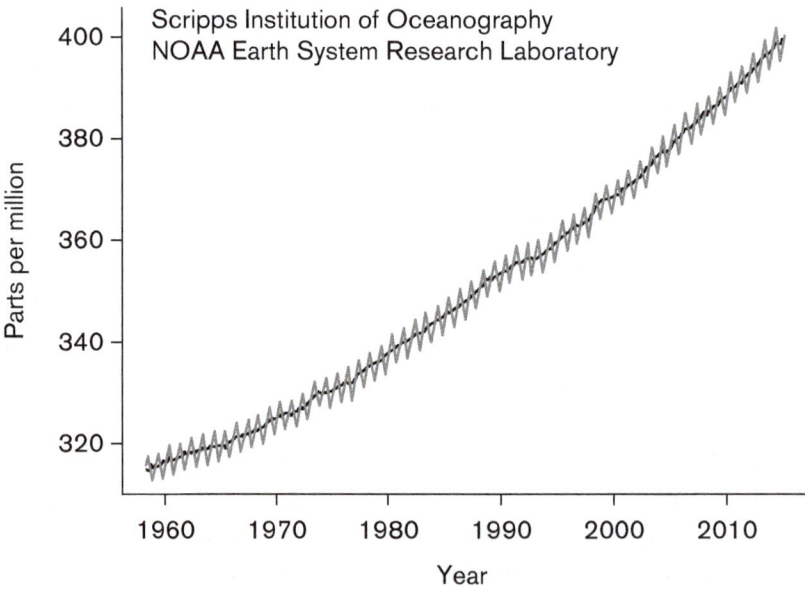

Figure 5.7 The carbon dioxide content of the atmosphere has been measured from the summit of Moana Loa in Hawai'i since the 1950s and has continued to increase. (Courtesy of Scripps Institution of Oceanography and NOAA).

Past and Present Measurements of Sea-Level Rise

While we cannot directly measure where sea level was 100,000 or 1,000,000 years ago, there is a direct connection between global climate and sea level, so that proxies, or records of past climate, can be used to determine prior sea levels. These records include the fossils preserved in deep-sea sediments (which extend back over 60 million years), ice cores from Antarctica and Greenland (extending back over 800,000 years), tree rings (which extend back about 5,000 years in the case of California's bristlecone pines), as well as pollen records from lake sediments and growth rings in deep-sea corals. We have also collected seafloor samples from the continental shelves around the world, which have preserved the record of the past 18,000 years of sea-level rise. On the floor of the North Sea, for example, between the United Kingdom and France, peat samples have been dredged up and dated using carbon-14. Peat forms in freshwater bogs or swamps, often in coastal areas very close to sea level. If we know the depth below sea level where the peat was recovered and can determine its age through radiocarbon dating, we can place a point on the sea-level rise curve of approximately where the shoreline or sea level was at that point in time. Datable samples of fossil corals can also help us determine where sea level was at various times in the past. Many samples from continental shelves around the world have enabled us to develop a good record of sea level for this entire time period (see Figure 1.1).

From the end of the last ice age approximately 18,000 years ago to about 7,000 years ago, sea level rose globally at an average rate of about 12 millimeters per year (mm/yr.), or 39 inches per century. Within this period of relatively rapid rise, however, there appear to have been pulses when rates were even higher for shorter intervals, 20 mm/yr., or 78 in./century, which are believed to have been due to major pulses of glacial melt from Antarctica. This is a very high rate relative to anything we have seen throughout the entire period of human history. Climate change slowed to a crawl about 7,000 years ago, and from geologic records of nearshore deposits recovered from the seafloor, it appears as though sea level rose at a very low rate, perhaps one millimeter per year, until the last 150 years or so.

The oldest tide gauges or coastal water level recorders were established in the mid-nineteenth century, and over the subsequent 150 years many others have been established along the world's coastlines. These gauges have continued to record the daily ranges but also long-term changes in sea level at each location, and a number of these now have histories of 75 to 100 years, which reveal a global average rate of sea-level rise for the past century of about 1.7 mm/yr. (~ 7 in./century). There is one major limitation to these tide gauge records, however. Because they are each anchored to some rigid mass of land, or some solid structure such as a bridge or wharf attached to land, they reflect the rise of elevation of the ocean relative to the land they are attached to.

In far northern latitudes where continents were depressed by thick covers of ice during the ice ages, there are many regions (e.g., Alaska, northern Canada, and Scandinavia) where the land is still rebounding from the ice removal (Figure 5.8). Think about depressing a mattress when you sit on it and then how it rebounds or recovers when you stand up, only the rebounding of a landmass takes

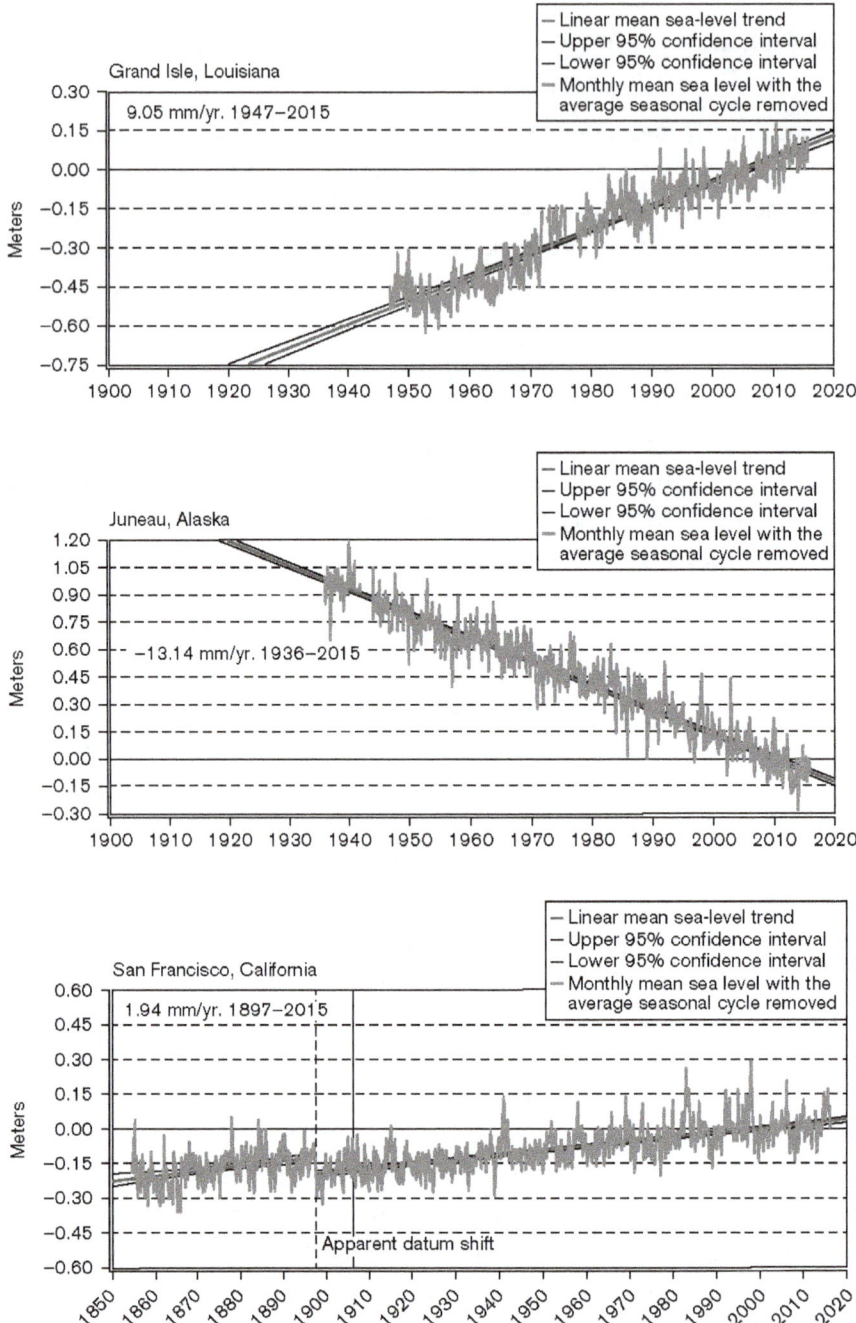

Figure 5.8 Sea-level rise values measured at tide gauges document local sea level relative to the landmass mass they are attached to. Alaska, San Francisco, and Louisiana tide gauge records illustrate these regional differences. (Courtesy of NOAA).

place very, very slowly. The land in these formerly glaciated northern latitudes is rising faster than the ocean such that sea level is actually dropping relative to the coastline. At Yakutat in southeastern Alaska, for example, sea level is falling at 17.6 mm/yr. (5.8 ft./century). Needless to say, the residents of Yakutat are not losing a lot of sleep over sea-level rise. The gauge in Furuogrund, Sweden, in the Gulf of Bothnia, shows a regular drop in sea level of 8.1 mm/yr. (2.7 ft./century) for the past hundred years. There are other areas scattered around the world, where the opposite is taking place. The coastline is subsiding, from either petroleum or groundwater withdrawals or from consolidation of organic-rich sediments, such that sea level is rising relative to the land at much higher than global rates. The Gulf Coast of the United States is a good example; at Eugene Isle, Louisiana, not far from New Orleans, sea level is rising at 9.6 mm/yr. (38 in./century; Figure 5.8), over five times the global average. The Gulf Coast has serious challenges ahead with future sea-level rise. Venice, Italy, is perhaps the most iconic of all the sinking regions around the world: because of land subsidence, sea level is rising there at nearly double the global average (see Figure 11.1).

These wide variations in relative sea-level measurements from fixed tide gauges scattered around the world's coastlines and the uncertainties about the accuracy of an average value due to the uneven geographic distribution of the gauges led to the implementation of a higher tech approach in the early 1990s. Several satellites were launched that use radar combined with GPS (Global Positioning Systems) to very precisely locate the positions of the satellites and their distance above the ocean surface in a process known as satellite altimetry. From 1993 to 2016 the absolute rate of global sea-level rise determined from satellite altimetry has been calculated at ~3.3 (± 0.4) mm/yr., almost double the 1.7 mm/yr. for the previous century based on the averaging of global tide gauges (Figure 5.9). This equates to 33 centimeters, or about

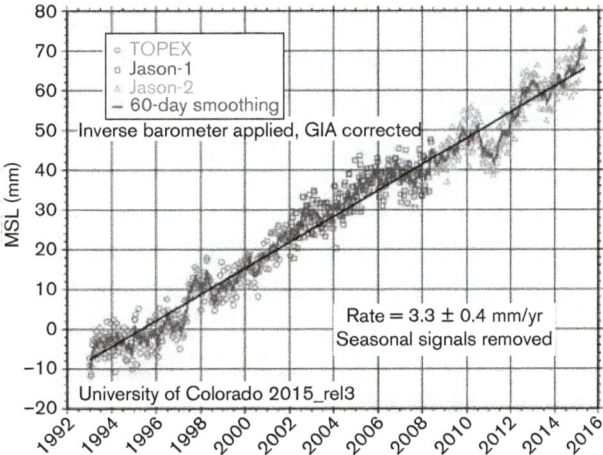

Figure 5.9 Global sea-level rise has been measured using satellite altimetry since 1993 and shows an average increase of about 3.3 mm per year. (Image: Steve Nerem, University of Colorado).

13 inches per century, and provides all coastal states and nations with a value that they can begin to plan around. It is also important for future coastal land use planning that regional tide gauges, if present, also be considered when there is significant tectonic activity and the land is either rising or sinking.

Global Warming and Future Sea-Level Projections

Over the past decade or so, the discussion of climate change and global warming has moved from the university seminar room and scientific conferences to city council chambers and the halls of Congress. For many, it initially seemed inconceivable that anything humans were doing could possibly have any significant effect on the atmosphere and temperature of the entire planet. Today, however, 97 percent of the climate science community believes that human activity has had a significant impact on climate change and the warming of the planet and its oceans, which is leading to increasing rates of sea-level rise. The questions coastal communities, states, and nations all need to face, if they haven't already, is what the future rate of sea-level rise will be, how it will affect their own region or jurisdiction, and how they are going to respond or adapt to the flooding, inundation, and shoreline retreat that will accompany the rise. In the United States, projecting population growth to 2100, there will be 4.2 million people living within 3 vertical feet of high tide. If sea level rises 6 vertical feet, the number of affected people rises to 13.1 million. With approximately 150 million people around the world today living within 3 feet of present sea level, for much of the population in large cities scattered around the world's coastlines, knowing how high sea level will be at various times in the future becomes an extremely important bit of information to know and use. Denial is not an effective strategy.

There are some coastal communities and areas that are already experiencing severe problems related to climate change and sea-level rise. One well-publicized group is the communities and island nations in the Pacific and Indian Oceans that occupy coral atolls, in most places just a few feet above sea level. The Maldives, Kiribati, Tarawa, and Tuvalu are just a few of the places where flooding and contamination of limited freshwater aquifers from a rising sea are endangering their very survival as communities and nations (Figure 5.10). A foot of sea-level rise can make a difference when you live almost at sea level to begin with and there is no higher ground to move to.

At the opposite climate extreme are the Alaskan Inupiaq and Yu'pik villages along the Beaufort, Chukchi, and Bering Seas. Kivalina has become the poster child for these subsistence hunting villages, which typically are each home to perhaps three hundred to six hundred people. Kivalina is on a sand and gravel barrier island, historically underlain by permafrost, which remained relatively stable. Climate change–related processes are now threatening Kivalina and other similar villages. Although still fairly stable as far as barrier islands go, the formerly solid permafrost foundation has now begin to thaw seasonally, such that it is much more vulnerable

Figure 5.10   This household on South Tarawa in the tropical Pacific is threatened by a retreating shoreline. Moving away from the shoreline is not an option as the island is very narrow and already crowded. (Photo: Government of Kiribati licensed under CC BY 3.0 via Wikimedia).

to wave attack (Figure 5.11). In addition, as the previous shore-fast ice, which seasonally armored the edge of the island, has melted and retreated each summer, there is more sea surface exposed to the wind, so larger waves can form. When the waves reach the shoreline there is no ice left to buffer their impact on the now weakened and thawing permafrost. In addition, thawing permafrost is more likely to lead to subsidence-related coastal flood hazards under increasing exposure to ice-free conditions. Each of these small villages is now facing a dilemma similar to that of the low-lying tropical Pacific islands.

Projecting or predicting future sea levels is a difficult problem for many reasons. As a distinguished scientist once said, "Prediction is difficult, especially about the future." Yet there are scientists from many different nations, many of them directly involved with the United Nations Intergovernmental Panel on Climate Change (IPCC), who are using all of the tools in their toolboxes to make the best possible predictions. Projections of future global sea level are commonly made using mathematical models of the fundamental processes that contribute to global sea-level change, primarily the transfer of freshwater from melting ice (the *cryosphere*) to the oceans and changes in water density (typically labeled *steric* changes) arising mainly from the thermal expansion of ocean water as it warms. At present we believe that ocean warming is contributing about one-third of the rise, with ice melt making up the remaining two-thirds. The former is more straightforward because we know the thermal expansion of water at different

Figure 5.11   The low bluffs of Kivalina, Alaska, are increasingly susceptible to subsidence and erosion as the underlying permafrost thaws. (Photo: Nicole Kinsman, NOAA/NOS).

temperatures, but it does involve some assumptions about how much individual layers or depths of the oceans' water will be warmed at various times in the future. The question of how much the melting of glaciers and ice sheets, primarily in Greenland and Antarctica, will contribute in the decades and centuries ahead, however, is much more complicated simply because as humans we have never before witnessed ice melt on this scale.

There is another approach to projecting future sea level, the semi-empirical approach, which is based on the observed relationship between past sea levels and past global temperatures. This approach essentially combines all of the individual contributors and their importance into a single correlation between global temperature and ocean level. There are geologic records from deep-sea sediments and ice cores, from coastlines and the continental shelves of the world, which provide good indicators of past sea level and global temperatures. These have been compared to give us a reasonable connection between future temperatures and how much sea-level rise they may generate.

Virtually all future projections include ranges based on different greenhouse gas emission scenarios, from aggressive reduction of emissions to business as usual. The range of estimates using these methods and the available data diverge as we look further into the future, simply because the uncertainties regarding global greenhouse gas emissions get larger. The largest uncertainties are not with the science, however, but with the politics and social responses to climate change in the 194 nations around the planet, primarily the top six greenhouse gas emitters, China, the United States, the European Union countries, India, Russia, and Japan.

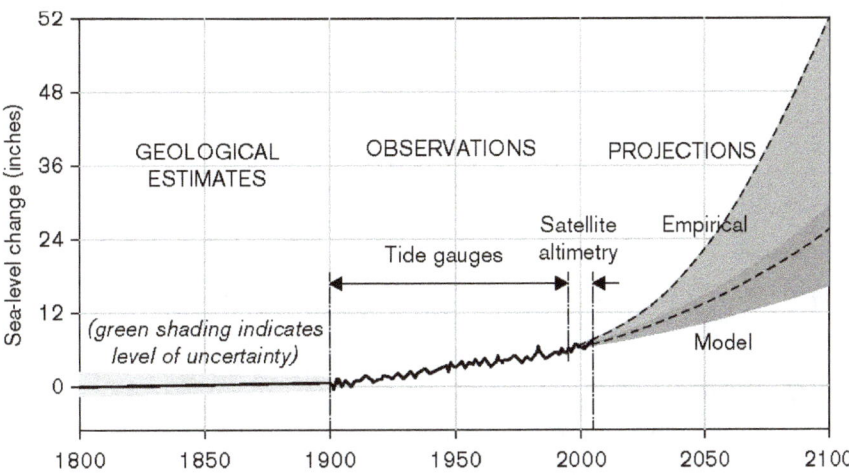

Figure 5.12 Past and present sea-level values and projections for the future out to 2100. (NAS-NRC Sea-Level Rise for the Coasts of California, Oregon and Washington).

While new projections and revisions continue to appear, a 2012 U.S. National Academy of Sciences report is still being used as a reference guide for the western states (California, Oregon, and Washington; Figure 5.12). The range of estimates for global sea-level rise from that report for 2050 lie between about 6 and 18 inches (15 and 45 cm), with a midpoint of about 12 inches (30 cm). For 2100, most ranges cluster between 20 and 55 inches (50 and 140 cm), with an average or midpoint of about 36 inches (90 cm). However, this will continue to be a moving target as long as greenhouse gas concentrations continue to increase and global warming persists, which is likely to be for at least several hundred years.

Assessing Future Vulnerability to Sea-Level Rise

Projections of future sea levels should be used as reasonable guides for what a community or region may expect in the future, but there is little doubt that these values will continue to be revised as new data become available. Any community, state, or nation planning for specific projects or proposals or developing long-term responses or adaptation strategies should also consider several factors:

1 the sea-level rise value from the closest tide gauge, because there are regional differences due to land motion;

2 the cost and lifetime of the facility being considered, whether a multimillion-dollar sewage treatment plant engineered to last fifty years or a city park or bike path;

3 the impact of damage to or loss of the project or facility and the cost of replacing or rebuilding it.

The impact of sea-level rise or flooding on a large power plant or an international airport (Figure 5.13), for example, will be of far greater consequence than damage or loss of a coastal parking lot or roadway, and for the former, a much more conservative value of future sea level should be used for planning and siting.

Figure 5.13 Areas around central San Francisco Bay that would be inundated by a 16-inch and a 55-inch rise in sea level. (Courtesy of San Francisco Bay Conservation and Development Commission).

While we do need to think in the long term and about what the range of future sea-level rise projections are for any particular area of coastline, there are short-term water levels that are of much greater concern. Some of these have been discussed in chapters 2, 3, and 4, events like tsunamis, hurricanes, severe storms, and extreme high tides, but there are the more annoying and regular events that are becoming more frequent, especially in some of the low-lying Atlantic seaboard cities. Norfolk, Atlantic City, Charleston, and Boston are just four of the places where sea levels rose 5 to 9 inches (12 to 22 cm) between 1970 and 2012, while the global value was only about 3 inches (7.5 cm). In addition, as mentioned earlier, these cities now experience between six and almost thirty days each year of nuisance flooding, when seawater at high tide is inundating streets and neighborhoods with increasing regularity (see Figure 5.4). Other Atlantic Coast cities, Wilmington, Washington, DC, and Annapolis, for example, are expected to experience approximately 130, 160, and 180 days each year, respectively, of nuisance flooding by 2030.

Along the California coast, "King tide" has become a common phrase in many low-lying communities. Several times each year, normally during the winter months, the irregular orbits of the Earth around the sun and the moon around the Earth conspire to produce these extreme high tides. In low-relief areas already very close to sea level, even a few inches can make a difference, especially when large storm waves arrive coincident with these extreme tides, giving us a glimpse of the future. The public has now gotten involved in documenting and publicizing these events on a social media website where images of flooding and very high tides are instantly uploaded, giving people a realistic image of what we are already experiencing in low-lying areas (Figure 5.14).

Figure 5.14 The Embarcadero in San Francisco during a very high (King) tide. (Photo: Sergio Ruiz © 2012).

Similar short-term causes of elevated sea levels are the periodic El Niño events, which affect the entire U.S. West Coast. During the large 1982–83 and 1997–98 El Niño events, the bulge of warm equatorial water that migrated up the West Coast elevated sea levels 12 to 24 inches (30 to 60 cm) above predicted tidal elevations. When combined with normal winter high tides and large waves, coastal flooding and erosion were severe. The 1982–83 El Niño inflicted over $240 million (in 2015 dollars) to oceanfront property in California when waves attacked beach-level or back-beach homes in places like Stinson Beach, Rio Del Mar, Aptos, Solimar, Faria, Malibu, Del Mar, Oceanside, and Imperial Beach. Damage was not restricted to broken windows and flooding of low-lying areas; thirty-three oceanfront homes were totally destroyed, and over three thousand homes and businesses were damaged, along with park improvements, roads, and other public infrastructure (Figure 5.15).

Figure 5.15 The 1983 El Niño damaged 3,000 oceanfront homes and businesses along the California coast and destroyed 33 homes. (Photo: Gary Griggs © 1983).

Where Do We Go From Here? Responding and Adapting

Sea-level rise has been recognized as a major threat to low-lying areas around the world since the issue of human-influenced global climate change first emerged in the 1980s. A growing number of reports and studies have illuminated and detailed the potential effects of a continuing and increased rate of sea-level rise. Many regions are already feeling the impacts, and there are many more that will in the decades ahead. Higher sea levels will erode beaches, dunes, bluffs, and cliffs and will temporarily flood and then permanently inundate low-relief estuaries, marshes, lagoons, and wetlands as well as any associated development and infrastructure. The effects of a rising sea on coastal cities and nations across the planet may be the biggest threat that human civilization has ever faced. Some small island nations that are only a few feet above sea level are already feeling the impacts and have been making plans for relocation to safer sites. The bad news and the good news is that it is still happening somewhat slowly and has been overshadowed by the more extreme short-term events. But the rising sea is a ramp, with these extreme events, whether storm waves, El Niños, or hurricanes, all riding on top.

Coastal communities, cities, states, and nations must start planning now, if they haven't already begun, for how they are going to respond to the future rise in sea level, which is now guaranteed and unavoidable. The coastal environments present in any region (whether beach, wetlands, estuaries, dunes, bluffs, or coral atolls) will respond differently, but assessing the vulnerabilities of each area to future sea-level rise is an important starting point. What are the historical or documented rates of bluff, dune, or shoreline retreat? Are there accurate elevation maps so that we not only know which areas have been flooded in the past and how often, but also which ones are likely to be affected in the future? This is the type of information that is needed to begin to develop response or adaptation plans.

The options available are limited, and many of them should be recognized as relatively short-term fixes. Beach nourishment and armoring, whether seawalls or rock revetments, fall into that category. While they may work for the immediate future (i.e., the next several decades), depending on both the rate of sea-level rise and the frequency and magnitude of extreme events, we need to accept the reality that they are not going to be successful over the long term. Managed retreat, or the gradual and planned retreat from the shoreline, will almost certainly become the accepted and only feasible long-term solution by midcentury and beyond. This is not going to be popular anywhere, but all protection ends somewhere, and we need to begin to look beyond the next election cycle and how we are going to live with the inevitable rise in sea level.

Discussion Questions

1   Why is coastal flooding becoming more common, even during sunny, calm days? When is this most likely to occur? Which U. S. cities are expected to have more than 100 such flood events per year by 2030?

2   Although sea level is rising at a global average of 3.3 mm/yr, this rate varies widely along different coastal locations. Why? Do an online search to discover where the rates are highest for the United States and globally.

3   Where is the damage from sea level rise already obvious? What additional concerns do you have about its continuation?

4   How should coastal cities and smaller communities adapt to rising sea levels? Where should the funding come from to do so?

CHAPTER 6

Selections from "Climate-Change Effects, Adaptation, and Mitigation"

John T. Abatzoglou, Crystal A. Kolden, Joseph F. C. DiMento,
Pamela Doughman, and Stefano Nespor

Modeling of climate provides information about the physical manifestation of climate change at many levels, from the global to the local. In order to understand these changes, additional steps are needed to translate physical changes into climate effects and to assess risk avoidance through mitigation and adaptation efforts.

Overall Global Consequences of Climate Change

The Intergovernmental Panel on Climate Change (IPCC) Fifth Assessment report released in 2013 provided strengthened support of widespread observed changes in global climate. Decadal global mean surface temperatures have progressively increased over the last three decades and were warmer than any decade since 1850 (IPCC 2013, SPM-3). The last month with global mean temperature below their twentieth-century average occurred in February 1985. Decreases in the mass of Greenland and Antarctic ice sheets, declines in sea-ice extent, and a contraction of Northern Hemisphere spring snow cover further substantiate a warming planet. The report further established clear links between human influences and many of the observed changes in climate.

These statements were released in spite of the relative lull of increases in global mean temperature over the past fifteen years. However, short-term records of global mean surface temperature are not a good proxy for assessing man-made changes in climate due to the natural year-to-year variability in climate. Dr. Richard Muller, who compiled an independent record of global temperature, described the fallacies of short-term records of climate change with the analog: "When walking up stairs in a tall building, it is a mistake to interpret a landing as the end of the climb" (2013). Indeed, the observational record of global mean temperature shows numerous perceived short-term "pauses" in warming that followed a continuation of long-term warming. With more warming already "in the pipeline" and atmospheric carbon dioxide concentrations escalating, the reality of a warming world and the consequences associated with it are increasingly apparent.

Global atmospheric carbon dioxide levels reached about 400 ppm in 2013. During 2012, CO_2 levels rose by 2.67 ppm over 2011 levels, according to the National Oceanic and Atmospheric Administration (NOAA). This increase represents the second-highest rise in CO_2 emission levels since record keeping began in 1959. Only 1998 showed a higher annual increase; during that year, CO_2 levels climbed by 2.93 ppm (Borenstein 2013). In its Fifth Assessment Report, the IPCC revealed that atmospheric CO_2 concentrations have increased by over 40 percent due to human activity since 1750, and by about 10 percent since 1990 (IPCC 2013, SPM-7).

In some scenarios, the IPCC (2007) estimated CO_2 concentrations may reach 600–1000 ppm by 2100. The IPCC estimated in 2007 that the ramp-up of atmospheric carbon dioxide could increase global average surface temperatures 1.8 to 4.0°C (3.2 to 7.2°F) by 2100 (Alley et al. 2007). In its Fifth Assessment, the IPCC put the prospective range for the end of the twenty-first century as "*likely* to exceed 1.5 degrees C relative to 1850 to 1900" under a number of sets of assumptions (emphasis in original; IPCC 2013, SPM-15). This may not seem like a major change; after all, day-to-day changes in temperature at any given point on the globe regularly exceed 4°C (7.2°F). However, a 4°C (7.2°F) increase in global average surface temperature may be better understood by considering that current-day temperatures are only about 6°C (11°F) warmer on average than the temperatures during the most recent ice age 20,000 years ago. Such a rise would represent a significant change.

Climate impacts arise in response to significant departures from what we consider a *normal* range of variability. These deviations from *normal* include both extreme weather events—for example, a landfalling hurricane like Hurricane Katrina and its effects on lives and infrastructure in Louisiana and Mississippi, or Superstorm Sandy and its impacts along the East Coast—and extreme climate events like the drought across the midwestern United States in the summer of 2012 and its far-reaching influences on agriculture and commodities. Extreme events are the main channel by which climate and society interact, and they attract the most climate-related media reports. Data for global natural catastrophes show there is an upward-sloping trend in the number of extreme weather events each year, with the frequency of storms, flooding, and related events growing more quickly than the frequency of extreme temperature, drought, and forest fires (Munich Re Group 2011, 11). The United States alone experienced fourteen weather disasters in 2011 and eleven in 2012 that each totaled over $1 billion in damages, with disasters encompassing hurricanes, tornadoes, wildfires, flooding, and drought (National Climatic Data Center 2013). However, the IPCC's Fifth Assessment Report indicates more uncertainty regarding the impact of climate change on hurricanes (Revkin 2013).

Extreme weather and climate shape society and ecosystems. For example, cities located in low-lying areas place restrictions on building in predefined flood plains, and infrastructure must be built to withstand a "one-hundred-year" flood. Ecosystems in California are drought-adapted, while those in northwestern Montana are cold-tolerant. The ability of society and ecosystems to adapt to climate change will be put to the test through extreme events. Extreme events are, by definition, low-risk high-impact events that are exceedingly rare in historic terms. Because climate change involves a shift in the statistics of weather, however, it is possible that even modest changes in climate will result in significant changes in the frequency and magnitude of extreme events (Meehl, Arblaster, and Tebaldi 2005; Diffenbaugh and Ashfaq 2010), which may be of greater importance than overall changes in the average climate. Containing the damage caused by climate change requires preparation for changes in the new normal and changes in the magnitude and frequency of extreme events.

What does a warmer world imply for human and ecological systems? This may be the most important question regarding climate change, as it may motivate mitigation and adaptation efforts. However, there is no straightforward answer. Effects can be detrimental and beneficial, direct and indirect. Impacts may occur through gradual changes in average climate as well as changes in extreme events. Regional and local changes in climate can be much more pronounced than globally averaged changes. To untangle some of these dynamics, the next section discusses expected consequences for human systems, followed by a discussion of expected consequences for natural systems.

Effects on Humans

The basic needs of human survival include food, water, and shelter. Climate enables the geographic existence of societies by providing resources to meet these needs. The ability of societies to sustain the essentials for a growing population during an era of changing climate is particularly challenged. Aside from climate change, natural variability in climate is an additional stressor in the network of factors (e.g., war, poverty, diseases) that affect the ability of populations to meet these needs. With climate change and a growing population, the challenge is more formidable, particularly for populations that struggle to meet the needs for survival and may not have resources to cope with an additional stressor.

Increased global population requires increased agricultural productivity. Despite increases in yield per acre as a result of mechanized agriculture, agricultural productivity is dependent on both available resources and a reliable climate. The projected impacts of climate change on agriculture are mixed. Increases in atmospheric carbon dioxide can act to stimulate crop productivity and increase water-use efficiency, albeit at the expense of nutritional quality. However, the direct effects of changes in precipitation and temperature are expected to alter agricultural productivity. Agriculture in areas with short growing seasons, such as Canada and Russia, where plants are currently growing below their optimum temperature, may benefit from a longer, warmer growing season. However, the 2010 drought and heat wave that resulted in a loss of over one-third of the Russian wheat crop hints there is a fine line between a longer growing season as a result of warming and potential heat and moisture stress that begets crop failure. Northward shifts in suitable climates are expected to push agriculture belts northward across the central United States, making the already marginal croplands of the US southern Great Plains unsustainable, and the prime cropland of the midwestern Great Plains less productive and profitable (e.g., Ortiz et al. 2008). Declining water availability in southern Asia, South America, and Africa will likely result in significant losses in major crops and threaten food sustainability (Lobell et al. 2008). Likewise, increases in drought frequency and magnitude will also threaten agriculture in southern Europe.

Climate scientists also expect that the *hydrologic cycle*—the cycling of water among the atmosphere, land, and oceans through precipitation and evaporation—will intensify in a warming planet. A warmer climate will enhance rates of evaporation and precipitation for the globe as a whole (Wetherald and Manabe 2002). Regional precipitation distributions may be drastically altered, leading to an increase in the intensity and frequency of rainfall in some regions and to pervasive drought in other regions. In addition to changes at the global level, local dynamics may influence intensity and frequency of precipitation as well (Seager et al. 2012).

The relationships between changes in the Arctic and climate change impacts are continuously being addressed; the dynamics are complex. Overall, as with many other phenomena in climate change, trends are consistent with major shifts despite periodic changes, for example, in the amount of ice that survives the summer melt (Overland et al. 2012; Samenow 2012; Samenow 2013; Gillis 2013a).

Changes in the Arctic due to climate change could cause weather systems to stall more frequently as they move across North America and Europe. Differences in temperatures between the Arctic and mid-latitudes affect the location and speed of the northern polar jet stream, which drives weather systems from west to east across the Northern Hemisphere. Because the Arctic is warming faster than the mid-latitudes, the difference in temperature is decreasing and the jet stream is slowing down. Also, the jet stream is developing large southward bends more frequently. The changes may increase the chances and duration of extreme weather in the Northern Hemisphere (Francis and Vavrus 2012).

Climate models predict a poleward shift in the storm track and descending branch of the Hadley circulation. The descending branch is currently near 30 degrees north and south of the equator, and is characterized by a downward movement in air that creates high pressure and limits cloud and precipitation for much of the year. Going forward, the climate models predict this downward flow of air will occur further northward in the Northern Hemisphere, thereby reducing precipitation across much of the southern tier of the United States and southern Europe. Increased evaporative losses due to rising temperatures are projected to increase aridity in the southwestern United States (Seager et al. 2007). This is expected to reduce runoff and reduce water availability for the southwestern United States, which may pose challenges for growing urban areas such as Las Vegas, Phoenix, and Los Angeles. Climate model simulations already suggest that some of the changes in the water cycle observed since 1950 can be attributed to anthropogenic influences. However, an increase in global mean precipitation is projected due to an intensified global water cycle in a warming planet, corroborating paleoclimatic studies that show that the Earth has been wetter during past warm epochs and cooler during past dry epochs. Despite an overall warmer and wetter planet, changes in precipitation are projected to be nonuniform. Significant increases in precipitation are projected at high latitudes and near the equator with decreases along the poleward edge of the subtropics near 30–35 degrees latitude (IPCC 2013, SPM-16). Thus, many regions in the arid and semi-arid mid-latitudes will likely experience less precipitation, while regions with typically abundant moisture will likely receive more precipitation (IPCC 2013, SPM-16). While the IPCC has high confidence in overall patterns of precipitation change, uncertainty remains regarding the magnitude of the changes.

Warming will result in additional challenges for water availability in mountainous areas where water is stored in the snowpack, a defining feature of states in the western United States. Observations have shown that a 1°C (1.8°F) warming has significantly altered the fraction of precipitation stored in mountain snowpack in the transitional elevations of the Sierra Nevada and Cascades since 1950 (Mote et al. 2005; Abatzoglou 2011). Apart from changes in precipitation, warming not only decreases mountain snowpack storage, but also changes the timing and magnitude of downstream water availability (Elsner et al. 2010; Cayan et al. 2008) and hydroelectric capacity (Hamlet et al. 2010). These changes—coupled with increased water demand associated with warmer temperatures, preexisting water allocations and rights, and inherent climate variability—are likely to make water an even more coveted resource, which may result in water conflicts in already stressed systems (Karl et al. 2009).

Some areas may see a decrease in water availability, with subsequent consequences for populations and the economy; however, other areas may be in for too much of a good thing. An intensification of the hydrologic cycle, coupled with an increase in the water-holding capacity of the atmosphere, has and will continue to increase the potential for extreme precipitation events (Min et al. 2011). Such changes not only increase potential flooding hazards but may also increase potential contamination of drinking water. For example, flooding in cities may overwhelm municipal water utilities, resulting in sewer overflow and standing water that is susceptible to high concentrations of *Giardia, Cryptosporidia,* and coliforms and vector-borne infections (as well as mosquito populations).

Tropical storms have been at the forefront of media attention after the record-setting summer of 2005 in the tropical Atlantic that featured fifteen hurricanes, four of which reached Category 5 status (winds exceeding 155 miles per hour). Although there appears to be no global trend in tropical storm frequency, the number of major hurricanes (Categories 4 and 5) has nearly doubled in the last thirty-five years (Webster et al. 2005). This observation is consistent with the increase in tropical surface ocean temperatures over the last fifty years. Although we cannot attribute a single hurricane or season to anthropogenic climate change, a warming of the tropical ocean would provide greater energy to fuel more powerful tropical storms and hurricanes.

In its Fifth Assessment the IPCC, after internal discussion regarding use of outlier scientific findings, predicted that sea levels will continue to rise during the twenty-first century. Under the different scenarios that the IPCC uses based on assumptions of various kinds, the rate of rise "will *very likely* exceed that observed during 1971–2010 due to increased ocean warming and increased loss of mass from glaciers and ice sheets" (emphasis in original; IPCC 2013, SPM-6). Over the period 1901–2010, global mean sea level rose by 0.19 (0.17 to 0.21) meters. (IPCC 2013, SPM-6). In its previous report, in 2007, the IPCC predicted that sea levels would increase 0.2 to 0.6 meters (0.92 to 1.40 feet) by 2100, compounding the effects of tropical storms on low-lying infrastructure (Alley et al. 2007). These estimates are considered very conservative. Scientists are looking to evidence from ancient coastlines to improve estimates of sea level rise due to climate change (Gillis 2013b).

As modeling of sea level rise matures, estimates of possible sea level rise by 2100 have been revised upward (Rahmstorf 2013a, 2013b). Informed by advances in sea level estimates, a 2012 NOAA study finds a greater than 9 in 10 chance that "global mean sea level will rise at least 0.2 meters (8 inches) and no more than 2.0 meters (6.6 feet) by 2100." To help decision makers reduce the vulnerability of coastal infrastructure while taking uncertainty into account, the study provides four scenarios. The study recommends the highest scenario, showing 2 meters of global sea level rise by 2100, for decisions on the construction of new power plants and other long-lived infrastructure. In addition, the study cautions that local changes in sea level will vary from place to place. In part, this is due to the fact that wind patterns and ocean temperatures vary around the planet, causing sea levels to rise in some areas more than in others (Emanuel 2012). In addition, sea level rise may vary due to local and regional conditions. For example, coastal areas are moving upward in Alaska and the Pacific Northwest, but subsiding in the Mississippi Delta (Parris et al. 2012, 1–3).

Sea level rise of 1 meter or more would displace hundreds of millions of people in low-lying coastal areas in Asia and inundate coastal cities, including London, Bangkok, Miami, New York, and Cairo. Low-lying areas in Florida and the US Gulf Coast would be particularly vulnerable to pressures from sea level rise, increased storm strength, and receding river deltas. The approximate 1-foot rise in sea level in New York since 1900 contributed to the record-setting storm surge experienced with Hurricane Sandy in October 2012. If current trends continue, members of human settlements affected by rising sea levels would be displaced and become "climate refugees."

The 2013 draft National Climate Assessment Development Advisory Committee (NCADAC) report warns that rising sea level, which is projected to rise by another 1 to 4 feet in the current century according to some models, and by as much as 6.6 feet by 2100 according to risk-based analyses, has the potential to negatively impact the nearly five million Americans who live within four feet of local high-tide levels. Coastal infrastructure—including roads, bridges, rail lines, energy infrastructure, and port facilities such as naval bases—is and will be increasingly at risk from storm surges, which are exacerbated by rising sea levels. Salt marshes, mangrove forests, barrier islands, and reefs defend coastal ecosystems and infrastructure, such as transit systems and buildings, against storm surges. The NCADAC report points out that climate change and human-caused ecosystem and landscape modifications often increase the vulnerability of these systems to damage from extreme events, while also reducing their natural capacity to mitigate the impacts of those events.

In addition, climate change models suggest there may be changes in the location and timing of *upwelling*. Upwelling provides nutrients for phytoplankton, which provide food for other marine life (Warner and Schofield 2012). Changes in the location and timing of upwelling of nutrients from ocean depths would cause changes in fisheries, which are already under pressure from overfishing in many locations. For low-lying island and coastal areas, changes in upwelling combined with sea level rise, ocean acidification, and vulnerability of coral reefs may create pressure for changes in maritime jurisdictional zones and other policies to better protect fisheries and livelihoods (Warner and Schofield 2012).

The direct effect of climate change on human health has been widely debated, as warming alone is projected to bring both detrimental and beneficial effects to humans. As heat-related mortality increases significantly when the air temperature exceeds 32°C (90°F) (Davis et al. 2003), warming is projected to result in heightened health risks for much of the world's population (Adger et al. 2007). Recent excessive heat waves, such as those that ravaged much of Europe in the summer of 2003 and western Russia in 2010, resulted in over 60,000 heat-related deaths. It is important to note that these heat waves may not have been directly caused by climate change, but rather by a persistent weather regime, although some recent work suggests a likely climate change link. Studies suggest that human-caused climate change loads the dice in favor of such events, making them far more likely (e.g., Stott, Stone, and Allen 2004; Rupp et al. 2012). Increases in temperature will allow heat waves to become more intense and longer-lasting (Diffenbaugh and Ashfaq 2010). Models also suggest that extremely warm summers, like the summer of 2003 across much of Europe, the summer of 2010 across western Russia, and the summer experienced across much of the United States in 2012, will be commonplace by the mid-twenty-first century. On the other hand, warming is projected to decrease cold-related mortality in northern Europe (Christidis et al. 2010). However, the world's least-developed countries lack the resources to cope with extreme temperature and are more susceptible to extreme heat. The IPCC has thus concluded that the net impacts of warming would increase human mortality.

Under such conditions, the 2013 draft NCADAC report cautions, new health threats will emerge and existing health threats will intensify. In addition, the report states that climate-change impacts "are expected to be most difficult for those with fewer resources to adapt. Some changes will be disruptive to society because our institutions and infrastructure have been designed for the relatively stable climate of the past, not the changing one of the present and future" (NCADAC 2013, draft).

Climate projections suggest that high latitudes will experience the greatest rate of warming over the next century. Warming at high latitudes may be beneficial for some communities, improving prospects for agriculture and timber harvests and ice-free shipping ports. However, the effects of warming will not be as kind for communities whose culture and way of life depend on ice. Changes in climate over the last few decades have already hurt Inuit villages. Many Alaskan coastal areas have experienced dwindling food supplies and infrastructure damages as reduced sea ice and melting permafrost have made villages vulnerable to erosion. As a result, coastal communities have relocated inland, while animals dependent on sea ice are in decline and are projected to experience continued declines on a warming planet (e.g., Durner et al. 2009).

The village of Newtok, Alaska, is surrounded on three sides by the Ninglick River as the river journeys to the Bering Sea. The Ninglick has steadily been consuming Newtok's land. From 1954 to 2003, erosion occurred at an average rate of 68 feet per year in the area of the village. (Climate Adaptation Knowledge Exchange 2010). The river and the erosion that is consuming Newtok's land have both been moving at extraordinary speed during recent years as a result of rapid, climate change-induced ice melt. A report by the US Army Corps of Engineers predicts that the highest point in the village might be underwater as early as 2017 and states it will not be possible to protect the village in its current location. Eventually, the village must be abandoned; the residents, whose forebears have been fishing and hunting in the region for centuries, will have to leave their ancestral lands, becoming America's first climate change refugees. If the village cannot be relocated, the 350-person community will be dispersed among the villages and towns of Alaska and to points unknown. The residents of Newtok are not alone: more than 180 native communities in Alaska are currently being flooded and losing their ancestral lands as the warming climate melts the Arctic ice (Goldenberg 2013).

Global warming's effects extend across the sectors of human health, the economy, politics, and international relations. Climate change has been projected to worsen urban air pollution (e.g., Hayhoe et al. 2004); increase malaria and other infectious diseases (Lafferty 2009); affect energy supply and demand (Miller et al. 2008; Sathaye et al. 2013) and transportation performance; and challenge regional and local economies. While some individuals and countries have resources to avoid many of the adverse effects of climate change, others may not be so resilient.

Wealthier nations and people may be able to lessen or adapt to the effects of climate change; however, the negative consequences of climate change fall disproportionately on poor and vulnerable nations and populations. According to the IPCC, the regional effects of climate change will vary over time and will be determined in part by the ability of different systems—both societal and environmental—to mitigate or adapt to climate change (NASA 2013).

Effects on Natural Systems

Ecosystems face numerous challenges brought about by climate change. Scientists can better understand and predict how future climate change is likely to impact natural systems by examining ecological responses to historical climate change using paleorecords such as tree rings and lake sediments. These repercussions take the form of both slow, long-term transitions in the biosphere and sudden, rapid ecological disturbances that radically shift the ecological make-up of local biota.

Marine ecosystems are starting to show the considerable negative effects of rising global ocean temperatures and ocean acidification. The ocean has been a sink of carbon dioxide, taking up approximately a third of the additional carbon emitted into the atmosphere since 1850. Over the last 250 years, oceans have absorbed 530 billion tons of CO_2, triggering a 30 percent increase in ocean acidity (National Resources Defense Council 2009).

This additional carbon translates into more dissolved carbon dioxide that alters ocean chemistry and makes it difficult for species to build shells and skeletons from calcium, especially in early life stages. Such changes affect oysters, some types of plankton, and other species important to marine food webs in coastal areas. Ocean acidification also slows photosynthesis in some species. It is still unclear to what extent key species in coastal and high-latitude marine ecosystems will be able to adapt to more acidic conditions, especially areas facing multiple difficulties such as warming waters, arrival of new predators, and pollution (National Research Council 2010). In addition, nearly half of the globe's coral reefs, home to the most biodiverse marine ecosystems on earth, are threatened by the combination of ocean acidification and rising temperatures. Research off the coast of Washington shows some organisms are adapting better than others, shifting the balance of predators and prey in the local ecosystem as ocean acidity increases (Wootton, Pfister, and Forester 2008).

The effects of climate change on terrestrial ecosystems are also of grave concern, especially those that provide considerable ecosystem services to humans and have additional aesthetic and spiritual values. Species generally fall into two categories: generalists, which are able to span broad ranges of climate conditions and are thus more resilient to changes in climate, and specialists, which generally exploit a very specific ecosystem niche and are often constrained geographically to specific areas. In light of these constraints, many of these specialists are adversely impacted by even small changes to the ecosystem, and are subsequently classified as threatened or endangered.

Paleorecords spanning thousands of years reveal that changes in vegetation composition across the landscape mirror changes in temperature, precipitation, and atmospheric composition. Long, slow changes in climate generally impact species by geographically altering what are called *bioclimatic envelopes*: the range of average temperature and precipitation conditions under which that species can continue to reproduce and sustain a viable community. It is widely held that warming will generally "push" the range of many species up in both latitude and elevation (e.g., Loarie et al. 2009), although changes in precipitation and moisture availability may drive more complex changes in species whose geographic range is more dependent upon moisture than on temperature (Crimmins et al. 2011, Williams et al. 2010, Lutz et al. 2010).

These shifts in species range are difficult for humans to observe as they take place over multiple decades, but changes in site-specific, local communities are more obvious. In Alaska, where significant warming has been observed over the past half-century, there has also been an expansion of the boreal forest to higher elevations, an increase in the density of shrubs on the tundra, and a drying pattern that has converted freshwater ponds and wetlands into dry, grassy meadows (Chapin et al. 2005, Riordan et al. 2006). Vegetation communities in any location naturally grow and change over time, in a process called *succession*. But climate change puts succession on a different trajectory, one that has a different end state, or "climax" stage. These two different vegetation community trajectories have different sets of faunal species associated with them, since the food chain starts at the lowest level with the herbivores that depend on specific plants for their diet. In the Alaska example, warming facilitates the uphill migration of the maximum tree line elevation and the expansion of the boreal forest habitat. Increases in shrub density on the tundra may provide more habitat for small mammals, but be less supportive for the large mammals with hooves (like caribou) that prefer low-to-moderate–density vegetation habitat. Meanwhile, the decrease in wetlands across the Alaskan interior reduces habitat available to the moose, beaver, and other water-loving species that forage or live in the wetlands.

Climate change alters the process of succession through two paths within the successional cycle of an ecosystem: competition and disturbance. There is competition for every available nutrient and resource on the landscape, and plants and animals evolve to take advantage of certain conditions in order to out-compete other species. Some reproduce at the highest rate, while others have a life cycle that capitalizes upon certain climatic conditions. For example, the spring bloom in the Northern Hemisphere has recently arrived between four and eight days earlier than it did during the 1950s due to warming (Schwartz, Ahas, and Aasa 2006). Differences in how species respond to climate change may potentially result in a mismatch in timing between consuming species and food resources. For example, changes in the extent of snow cover weaken the ability of snowshoe hares to camouflage themselves and avoid predation (Mills et al. 2013). Floral species that depend on animal and bird seed consumption and elimination in order to reproduce suffer a reduction in their reproductive potential due to the absence of the migrating fauna, while other species that require winds to disperse their seed may capitalize on an opportunity to expand. This type of pattern is mirrored in the case of changes in precipitation; some species capitalize on drought conditions, while others take advantage of wetter conditions. This sort of opportunism is one of the factors that encourage the spread of invasive species, which capitalize on "new" climate conditions as the existing species struggle to adapt. Buffelgrass (*Pennisetum ciliare*), a species native to much of subtropical Africa and Asia and introduced as a livestock forage in the southwestern United States and Mexico in the 1930s, has outcompeted native vegetation and expanded its range in recent years as a result of warming conditions and abnormal years of precipitation. Buffelgrass is highly flammable; as a result of its spread, wildfires have become increasingly prevalent within the fire-intolerant ecosystem across the Sonoran desert. Projected changes in climate are likely to further benefit the invasion of such annual grasses and threaten desert and rangeland ecosystems across the southwestern United States (Archer and Predick 2008).

The other path by which climate change affects ecological communities is through periodic disturbance. Ecological disturbances such as fire, flood, avalanches, insect infestations, severe weather events, disease, and regional die-off occur naturally in almost all ecosystems; these disturbance events "reset the clock" on succession processes and have occurred for millennia. Some species are not only adapted to disturbance, but require it to sustain the population, such as giant sequoia trees that only sprout when a low-intensity fire opens their cones. The frequency and intensity of these disturbance events, however, is dramatically altered by climate change, to the detriment of species that are less resilient or adaptable. Wildfires are a regular occurrence across much of the western United States, and most forest and grassland ecosystems there are adapted to regular wildfire events. However, the projected warmer and drier summers across regions like the northern Rocky Mountains by the middle portion of the twenty-first century will significantly increase the possibility of larger and more frequent fires (Brown, Hall, and Westerling 2004; Westerling et al. 2011b). This type of climate-induced change in the wildfire regime favors the expansion of species that are adapted to more frequent fires (like ponderosa pine trees) over the existing species that cannot resist fire and take longer to reestablish (such as lodgepole pines and spruce trees). In the southwestern US deserts, projected increased fire danger has the potential to combine with an ongoing expansion of annual invasive grasses in a self-perpetuating worsening cycle that will eventually convert deserts currently populated by cacti and dry habitat shrubs into deserts covered by a homogenous invasive grassland (Abatzoglou and Kolden 2011), effectively eliminating the habitat of the threatened desert tortoise (*Gopherus agassizii*) and other species. In addition to its impact on ecosystems, wildfire affects humans via its direct impact on life and property, regional impact on air quality and human health, and the significant economic costs of fire suppression.

Threatened and endangered species like the desert tortoise face the greatest potential consequences from climate change. The most vulnerable species reside in novel climates: geographic locations that harbor a special climate, ultimately providing a unique habitat exploited by a specialist species. Often, these specialists are *endemic*, meaning they are only found in a specific location. Climate change will extinguish these novel climates and push vulnerable and endemic species toward extinction in two ways: 1) by pushing the climatic conditions at that location beyond a survivable threshold for the vulnerable species, thereby destroying the novel climate (Williams et al. 2007); or 2) altering the characteristics of ecological disturbance so that the species is unable to survive and adapt to the new disturbance. In the Arctic, the polar bear population has been reduced through climate change–related habitat alterations to the point where the species is now listed as threatened. Climate change has significantly reduced the polar sea ice, and warmer autumn months in the last decade have delayed the formation of sea ice along the coast by several weeks, at a time when polar bears are at greatest risk. In a cycle that worsens over time in response to previous events, the warming also triggers stronger autumn storms that both degrade shoreline habitat for polar bears and their food sources, and make it even more difficult for sea ice to form. A 2007 US Geological Survey report (released prior to the record 2007 low sea-ice event) conservatively projected that polar bears will decline throughout all of their range (sea-ice habitat) during the twenty-first century and will be extinct within portions of their range within seventy years at the projected rate of warming (Amstrup, Marcot, and Douglas 2007).

Case Study: California

The state of California provides a suitable test bed for examining regional and local climate change repercussions because of the complex physical controls on regional climate across the state and the sensitivity of the regional economy to climate changes (Abatzoglou et al. 2009; Hayhoe et al. 2004). California is also the most populous US state and contributes about 6 percent of total US greenhouse gas emissions. Finally, it has exhibited a progressive stance on environmental policies pertaining to climate change, and is regarded as a leader to guide climate actions for other US states.

California has witnessed an increase in temperature of 0.8–1.5°C (1.4–2.4°F) over the last century, with warming trends generally more pronounced for minimum temperatures than maximum temperatures (Cordero et al. 2011). Results from a large number of modeling runs project that statewide temperatures will increase 1.5–4.5°C (2.5–8°F) over the course of the twenty-first century (Cayan et al. 2008). As of March 2013, the most recent set of climate models developed for the Fifth Assessment report of the IPCC concur with such projections. Models project the warming to be most pronounced during midsummer (up to 6.5°C or 11.7°F warmer), "enough to make many coastal cities feel like inland cities do today, and enough to make inland cities feel like Death Valley" (Hayhoe et al. 2004). However, other recent work suggests that climate change may increase coastal upwelling and lower temperatures near the California coast.

An increase in average temperatures means that heat waves will be more frequent and more intense. During July 2006, an extremely warm air mass stagnated over the western United States. Fresno experienced six consecutive days of 43°C (110°F) temperatures; the low temperature in Death Valley on July 24 was 38°C (100°F). An extended period of record-setting heat during the latter half of the month contributed to the death of over 160 people statewide (Gershunov et al. 2009). Models predict that heat waves of the same magnitude will be commonplace in California by the end of the twenty-first century (Miller et al. 2008). Moreover, future heat waves in California are projected to have significant impacts in coastal areas, where most of the state's population lives. The coastal areas exhibit more humid conditions that inhibit overnight cooling. High temperatures and humid conditions have significant negative consequences for human health (Gershunov and Guirguis 2012). Increases in temperature, in conjunction with warming's effect on air quality, could indirectly increase respiratory and cardiovascular problems that currently contribute to approximately 8,800 deaths and over a billion dollars in health care costs annually in California (Karl et al. 2009). However, there are steps we can take to avoid this outcome. If more stringent control measures are implemented, air quality may continue to improve in California, even with a warming atmosphere.

Predicted statewide increases in temperature and population will drive the demand for air conditioning and energy use, unless increased energy efficiency and self-generation can make up the difference. In California, the peak hour greatly exceeds the energy needs for the rest of the year. With expected increases in population and temperature, peak electricity demands may march upward and require changes in the electrical system to cope with additional stressors (Miller et al. 2008). However, the number of rooftop photovoltaic systems (which generate electricity from sunlight) is growing quickly, changing the shape of the net load curve for much of the state and increasing the need for fast-ramping energy resources (California Independent System Operator 2013).

Despite mixed projections of changes in annual precipitation totals for the state, warming will induce significant changes in hydrology as more precipitation in the Sierra Nevada falls as rain rather than snow and warmer temperatures accelerate snowmelt. Projections of the amount of water stored in snowpack on April 1st, which are often used to develop water supply projections for the summer, are expected to decline 32–79 percent by the late twenty-first century (Cayan et al. 2008). The trickle-down effect of changing water resources is likely to hurt the state's economy and vitality and damage a number of treasured ecological facets that define the state. First, warmer and wetter storms during winter are projected to increase winter runoff. Since present-day reservoirs cannot store anticipated increases in winter runoff for use during the dry season, peak floods will likely be more intense (Anderson et al. 2008, Das et al. 2011). Not only will flood impacts be worsened by rising sea levels, the combination of the two may result in salt-water intrusion to the Sacramento–San Joaquin delta, further threatening the state's water supply. Aside from such threats, operating costs to sustain water availability for the state under a warmer and drier scenario could increase nearly $500 million per year by the mid-twenty-first century (Medellin-Azuara et al. 2008).

A decline in water availability would damage California's multibillion-dollar agriculture industry. A combination of more winter floods, reduced water availability, and increased evapotranspiration (a combination of evaporation from water and soils and release of moisture from plants) during the growing season means that water-intensive crops are likely to suffer (Field et al. 1999; Wilkinson et al. 2002; Miller, Bashford, and Strem 2003; Cayan et al. 2011). Warmer temperatures and decreased soil moisture would dramatically increase irrigation demands and reduce crop yields. Warmer temperatures during winter may reduce the number of chill hours needed for fruit and nut trees to become dormant and set fruit, thereby limiting productivity in many areas (Baldocchi and Wong 2008). Likewise, the state's treasured wine industry is projected to be in jeopardy as warming across many world-class wine grape–growing areas would make them inhospitable to fine wine production (White et al. 2006, Diffenbaugh et al. 2011). However, the cooler climates of coastal California and the foothills of the Sierra Nevada may still be viable areas for wine grape production in a changing climate.

Climate change's ecological effects in California will be highly variable, since the state has arguably the greatest biological diversity in the contiguous United States, with numerous endemic and vulnerable species occupying hundreds of microclimate niches. Generally, however, climate change is expected to alter the timing of important ecological events, such as spring green-up and the onset of the bloom, while also shifting ecological zones northward and upward in elevation. Assuming that plants maintain ideal growing conditions under a temperature optimum, a 3°C (5.4°F) warming over the next century would require current vegetation distributions to migrate northward about 5 kilometers (3 miles) per decade (Loarie et al. 2009). The warming of waters in the Sacramento Delta is expected to increase the frequency of water temperatures lethal to delta smelt, an endemic species critical to ecosystem health (Cloern et al. 2011).

In the northwestern portion of the state, the consequences of warming and summer drying will reduce the viability of the temperate forest and make it susceptible to more frequent and more devastating wildfires (Westerling et al. 2011a). This is likely to help change the rainforest to a mixed-conifer forest not unlike that which carpets the eastern (lee) side of the coastal mountain range and the mid-elevation slopes of the Sierra Nevada. To the south, more frequent and severe wildfires have already been chipping away at the remnant stands of big-cone pine crowning the high peaks in the Big Sur region; these stands will disappear entirely and be replaced by shrubs. The range of coastal redwoods may shrink due to warming and increased moisture stress (Johnstone and Dawson 2010). Invasive species, which have already expanded across much of coastal California, will capitalize on opportunities to fill these empty niches, to the detriment of native species.

Along the 400-mile length of the Sierra Nevada, the species populating the high alpine meadows and forests will be faced with a reduced snowpack and longer snow-free season. This may be an opportunity for some species of animals and birds. Other species, however, may experience die-off or increased competition as temperatures rise and moisture declines. A lengthening of summer moisture-limiting conditions will be detrimental to the array of wildflowers that are accustomed to ever-moist soils in the alpine meadows, and the bird and insect species that depend heavily upon this flora will see a reduced food source. Further down the slope, species may move depending on their primary requirement (temperature, moisture); the diverging movement will disrupt current plant community dynamics and favor those that can adapt quickly.

In the deserts of eastern California, the situation is already dire. Disturbance in deserts most often takes the form of rare and intense rainfall events that unleash floods across a landscape with little vegetation to hold it together. Invasive annual grasses have covered much of the Mojave Desert and northeastern California's Great Basin over the last two decades, replacing native xeric (dry habitat) shrubs such as black brush, creosote bush, and sagebrush. One of the primary drivers of this replacement has been increasing wildfire, which was relatively rare prior to European settlement. Now, however, wildfire occurs in a positive feedback loop with invasive grasses: occurrence of one increases the occurrence of the other. Desert species are highly specialized to deal with long periods of drought followed by short, infrequent rainfall events and do not quickly recover from more frequent disturbances. Nor do they adapt well to these types of rapid changes; there is considerable evidence that such iconic species as the Joshua tree, with its poor dispersion capacity, will disappear because of climate change (Cole et al. 2011).

Climate change–induced alterations will be particularly acute for refuges and corridors for particular species. The wildlife refuges that dot the Sacramento Valley as stopovers for several hundred bird species that migrate over the state each year may dry up, and no similar or equal location exists further north, effectively eliminating this refuge. Existing habitat corridors for migrating elk and mule deer in the Eastern Sierra are likely to be left empty of these large mammals, while they struggle to find new corridors further upslope and northward amid private, unregulated lands. In a state as populous and developed as California, species may be unable to move onward and upward, or into more suitable locations, to adapt to climate change. Millions of Californians have built the state into a network of development that creates a formidable roadblock to finding new, more suitable habitat.

The California coastal waters are rich in marine life attracted by the upwelling of nutrient-rich deep water (Bograd et al. 2009; NOAA 2012). The upwelling waters are also rich in CO_2. As a result, researchers are concerned that California waters may soon experience high levels of acidification. A study completed in 2012 suggests that ocean acidification will lead to a substantial drop in seawater calcium needed for shellfish, small marine snails, and other marine life in California coastal waters within the next twenty to thirty years. It is not clear how California marine life will respond to this change along with concurrent changes expected from warming ocean temperatures and reduced dissolved oxygen levels (Gruber et al. 2012). The oyster industry off the coast of California and neighboring states has already seen a steep decline linked to ocean acidification. Awareness of this link has led to opportunities for adaptation in the oyster industry (Weiss 2012).

Case Study: New York

The state of New York provides an additional case study and interesting contrast in climate change and its projected consequences on the opposite side of the country. Climate-change effects in New York in many instances will present different concerns, or different levels of concern, from those experienced in California. For example, while California is more threatened by drought and ocean acidification along its extensive shoreline, New York is more vulnerable to sea level rise, powerful rainstorms, and storm surges. While hurricanes and associated storm surges are of great concern to the people of New York City, hurricanes do not impact California and storm surges are of less concern along California's coastlines.

A report written by fifty scientists from Cornell University, Columbia University, and the City University of New York was commissioned and released by the New York State Energy Research and Development Authority. The report, which examined recent weather patterns in comparison with historical patterns, predicts that average annual temperatures in New York State will rise by 4°F to 9°F by 2080. Heat waves will become more frequent and intense, and the state's air quality will worsen (Rosenzweig et al. 2011).

A National Resources Defense Council (NRDC) report states that twenty-three counties in New York now experience unhealthy smog levels that will worsen in a warmer climate. Increased heat, particularly in combination with poor air quality, will result in increases in human heat-related illnesses and deaths. Heat-related mortality in the state is projected to rise by 70 percent by mid-century. In addition, ragweed pollution is currently a problem in thirty-seven counties in the state; climate change will result in an increase in pollen production by plants, an increase in the seasonal duration of various types of allergen production, and an associated increase in allergies and asthma (NRDC 2013).

Other sectors, such as energy production, will also be challenged or overwhelmed by the increase in heat. Electrical demand is expected to increase significantly during the warmer months as people try to combat the heat by cooling residences, businesses, and public buildings. (Rosenzweig et al. 2011).

The NRDC report also predicts that waterborne illnesses and infectious diseases like dengue fever, West Nile virus, and Lyme disease will increase. A warming climate will also result in toxic algal blooms in bodies of freshwater and along coastal shorelines (NRDC 2013); this will further threaten human health and will lead to the death of aquatic species as well.

Increasing winter temperatures may shorten the duration of Great Lakes ice cover, permitting more moisture to rise from the lakes. The additional moisture would then fall as snow in western New York during cold weather. While the severity of winter snowfalls is likely to increase, the number of days with snow on the ground is likely to decrease. In coastal regions and estuaries, nor'easter blizzards may more often turn into severe rainstorms. Precipitation is predicted to rise by as much as 5 to 15 percent and will often occur as heavy downpours, leading to flooding and its associated impact on water resource quality and agriculture, as well as infrastructure such as transportation systems, buildings, water treatment plants, sewer systems, and roadways and bridges. Much of the increase in precipitation is anticipated to occur during the winter months (Rosenzweig et al. 2011). The increase in flooding and anticipated increase in sewer system failures due to extreme weather events is expected to threaten human health in seventy counties within the state (NRDC 2013).

While the winter months will receive more precipitation, less rainfall is predicted for the summer months. As a result, summer droughts are expected to increase in duration and intensity. Drought-related impacts will further affect water supplies, agricultural production, natural ecosystems, ecosystem processes, and energy production in the state (Rosenzweig et al. 2011). Summer droughts are expected to result in water shortages in 26 percent of New York's counties by mid-century (NRDC 2013).

As the climate changes, plant, animal, reptile, and bird populations are expected to crash. Species will shift their ranges in response to changes in their habitats and food supplies, or will attempt to do so. Many species will be unable to make the transitions quickly enough, or will be unable to find enough suitable habitat in other locations. Invasive insects and weedy species are expected to increase. This will result in wide-ranging effects on natural systems as the changes cascade through interrelated species' food webs.

As stated in the state's 2010 Interim Climate Action Plan, New York is already experiencing warmer temperatures, especially in winter, leading to shorter periods of winter snow cover and a decline in winter sports and tourism. Climate change is likely to result in significant agricultural-sector economic costs in New York from decreased crop yields and increased heat stress–induced decline in dairy cow milk production (New York State Department of Environmental Conservation 2010).

In addition, climate-change effects on the state's sugar maples may even threaten the survival of this iconic resource (Rosenzweig et al. 2011). In addition to being the primary source of maple syrup, sugar maples are responsible for much of the beautiful fall foliage that attracts visitors from around the globe; if this resource is lost, fall tourism is also likely to suffer.

Changes in water quality and abundance will also lead to decreased recreational and commercial fishing (New York State Department of Environmental Conservation 2010). This will represent an additional impact to the tourism and sport fishing industries. Meanwhile, increases in sea level will result in the loss of coastal wetlands and the species that depend upon them during all or part of the life cycles. As the oceans continue to warm, higher water temperatures will lead to declines in fish and shellfish populations. Already, lobsters and some other cooler-water marine species are shifting their ranges northward and out of the state, while the warmer waters are favoring the increase in numbers of species like the blue claw crab (New York State Department of Environmental Conservation 2010). Sea level rise will also promote saltwater intrusion in some areas.

Some economic costs associated with climate change—while just as real—are more difficult to quantify. How do we put a price on the heritage value of alpine forests or their function as habitat for endangered species? Moreover, some costs cannot be accurately calculated due to the unpredictable nature of future changes. On the one hand, climate change in other parts of the state, including in the Snow Belt, may bring about downturns in snow-based tourism. On the other, increased economic opportunity may arise as a result of a more moderate upstate climate and the additional availability of water resources. As Ackerman (2009, 54) states, "There is no hope of coming up with a single dollar amount that adequately summarizes the full range of climate impacts; too many of the impacts are incapable of being measured in monetary terms. Yet even without an impossibly comprehensive summary number, there are ample grounds for taking action to reduce climate damages."

In 2013, Governor Cuomo and the state of New York began to caution bond investors that climate change may pose a significant long-term risk to the state's financial well-being. Citing Hurricane Sandy and Tropical Storms Irene and Lee, the caution to investors is included on the bonds along with warnings about additional risks that could adversely impact the state's finances, including federal spending cuts, litigation against the state, and potential labor negotiations. According to Richard Azzopardi, a spokesman for the governor's office, "The state determined that observed effects of climate change, such as rising sea levels, and potential effects of climate change, such as the frequency and intensity of storms, presented economic and financial risks to the state.... The extreme weather events of the last two years highlighted real and potential costs from extreme weather events, including the need to harden the state's infrastructure and improve disaster preparedness, both of which have been a priority of the governor" (Kaplan 2013).

The Interim Climate Action Plan (New York State Department of Environmental Conservation 2010) reports that global sea level rise, compounded by local subsidence of coastal land, have resulted in a local increase in sea level of 3 cm (1.2 inches) per decade in New York. By 2100, relative to the 2000–2004 baseline, sea levels are expected to rise anywhere between about 30 cm and 58 cm (12 and 23 inches) in the New York City coastal areas and along the Hudson River estuary. When rapid glacial ice melt is factored into the models, sea levels are projected to rise almost 140 cm (55 inches) in the Lower Hudson Valley, New York City, and Long Island. Sea level rise will immensely exacerbate existing risks to coastal populations and lead to permanent inundation of low-lying areas, increased coastal and beach erosion, an increase in the frequency of storm surge–related flooding, and contamination of the Hudson River, coastal estuaries, and groundwater-based freshwater supplies with salt water. High water levels, heavy precipitation, and strong winds from fierce coastal storms already result in billions of dollars in damages, disrupt transportation and power production or distribution systems, and dramatically alter the barrier islands that help to protect New York City and the coastline from powerful storms and storm surges. Rising sea levels will worsen storm surge-related flooding, threaten vulnerable energy production and telecommunication facilities in coastal areas, and inundate coastal marshes, wetlands, and estuaries.

Judging from the impacts of Hurricane Sandy in 2012, which caused billions of dollars of damage (FEMA 2013), a sea level rise of even 23 inches could be devastating for the city of New York, damaging or destroying natural areas, buildings, roadways and bridges, rail and subway lines, tunnels into and off the island of Manhattan, utility and telecommunications networks, wastewater treatment facilities, recreational facilities, and other forms of infrastructure. A 55-inch increase in sea level could inundate large swaths of the city and leave noninundated areas of the city increasingly vulnerable to coastal storms and storm surges.

Sea level rise, higher temperatures, extreme weather events, and increases in precipitation, flooding, and coastal erosion will put transportation infrastructure and operations at risk. Asphalt pavements; other road, bridge, and runway surfaces; railroad tracks; and electrical wiring and conduits may be damaged by more frequent and extreme temperatures and by intense storms. Air conditioning requirements in transportation vehicles and in tunnels will increase, placing greater demand on energy systems. High winds during intense storms may result in more frequent temporary closures or restricted use of larger bridges (Rosenzweig et al. 2011).

Flooding, washouts and erosion, and mudslides or landslides will also adversely impact roadways and railroads running along inland rivers and streams, in areas where existing drainage is insufficient to cope with intense storms, or in areas of steep terrain. All transportation systems in the coastal regions are threatened by coastal storms and related storm-surge flooding hazards. In New York, some of these systems are located at low elevations along bodies of water. Some subways, railroads, and highways tunnel beneath the city below sea level. Sea level rise, extreme storm events, and storm surges will dramatically increase the probability of flooding. Sea level rise will eventually inundate low-lying transportation systems permanently unless costly mitigation or adaptation measures are taken. The increase in storm intensity will affect air transportation, increasing the number of delayed or canceled flights, temporarily shutting down airports, and detouring flights to other airports. Because hotter air provides less lift, airport runways may have to increase in length (Rosenzweig et al. 2011).

Intense storms increase and redistribute sediment loads in rivers, harbors, and shipping lanes, potentially increasing the need for dredging operations. Bridge foundations in some rivers are also scoured by *sediment transport* (the movement of various-sized particles due to the combined effects of gravity and the movement of the fluid in which the particles are located) during strong storms and heavy water flows. Yet sediment transport in New York waterways is not clearly understood, and sediment transport under future climate conditions is even less well understood. Sea level rise may dominate over time, reducing the need for dredging operations. On the other hand, sea level rise may also reduce bridge clearances over waterways below the limits set by the US Coast Guard or by other jurisdictions (Rosenzweig et al. 2011).

Increased evaporation in the Great Lakes or along the St. Lawrence River Seaway under severe and prolonged droughts and extended heat waves is likely to lower water levels enough to impede shipping. However, a warming climate should prolong the ice-free shipping season on the lakes and seaway, making them less prone to the ice floes or shore-to-shore freezes that have occurred in the past (Rosenzweig et al. 2011).

For the telecommunications sector, the greatest threats posed by climate change will result from an increase in short-duration but intense downpours, or large-scale weather events such as hurricanes that bring high winds, lightning, and flooding. These types of events are predicted to increase statewide. Nor'easters and winter precipitation such as freezing rain, ice, and heavy snow also impact telecommunication systems. As noted above, these events may lessen in many parts of the state and may worsen in areas subject to lake-effect weather systems (Rosenzweig et al. 2011). [...]

Impact Avoidance: Adaptation, Mitigation, and Geoengineering

Future changes in climate have been portrayed as cataclysmic, which results in an abstractness that is difficult to deal with. What can we do in the face of such predictions? We can accept the fate of dire outcomes, or devise ways to counter the detrimental effects of change. Three fundamental means have been proposed to temper detrimental impacts: adaptation, mitigation, and geoengineering.

Adaptation involves using climate projections to reduce risks associated with a changing climate. For example, low-lying areas vulnerable to sea level rise are developing strategies to minimize socioeconomic loss (Frazier et al. 2010). Another strategy involves managing fresh-water resources, particularly in regions vulnerable to water shortages (e.g., Roy et al. 2010). In addition to precipitation changes, warming in snow-dominated areas is likely to alter the seasonal availability of water and force water managers to develop new approaches to optimize water resources (e.g., Milly et al. 2008). Likewise, adaptation strategies are being developed for agriculture (Lobell et al. 2008), species habitats and refuges (Rehfeldt et al. 2006), and forest disturbances (Littell et al. 2010).

Monitoring local variations in ocean acidity is making it possible for the oyster industry in the Pacific Northwest to adapt to ocean acidification. Another adaptation strategy is to protect and plant seagrass meadows and other marine plants to help offset ocean acidification locally by uptaking CO_2 from the seawater. A third strategy is to reduce other stressors, such as pollution, to help increase the resilience of coastal marine life at the local level (Ocean Carbon and Biogeochemistry Program, Ocean Acidification Subcommittee 2012).

The United Nations Environmental Programme (UNEP) has launched the Blue Carbon Initiative to protect coastal and marine mangrove, salt marsh, and seagrass ecosystems that provide carbon sequestration and other important ecological services (UNEP 2012). Protecting these systems helps local areas adapt to rising ocean acidity and helps mitigate global levels of anthropogenic greenhouse gases.

In its discussion on adapting to the impacts of climate change, the draft 2013 NCADAC report concludes:

> Because of the influence of human activities, the past climate is no longer a sufficient indicator of future conditions. Planning and managing based on the climate of the last century means that tolerances of some infrastructure and species will be exceeded. For example, building codes and landscaping ordinances will likely need to be updated not only for energy efficiency, but also to conserve water supplies, protect against insects that spread disease, reduce susceptibility to heat stress, and improve protection against extreme events. ... Adaptation considerations include local,

state, regional, national, and international jurisdictional issues. ... Both "bottom up" community planning and "top down" national strategies may help regions deal with impacts such as increases in electrical brownouts, heat stress, floods, and wildfires. Such a mix of approaches will require cross-boundary coordination at multiple levels as operational agencies integrate adaptation planning into their programs (NCADAC 2013 executive summary, 6).

Mitigation involves efforts to reduce the amount of carbon dioxide in the atmosphere. Some anthropogenic change is already inevitable due to the long lifetimes of greenhouse gases and because ocean waters and feedback processes retain heat for long periods of time. Even if anthropogenic emissions stopped today, the planet would continue to warm for the next twenty to fifty years. Reducing further emissions will reduce the magnitude of climate change, the scope and scale of damaging impacts, and the costs of adaptation.

Efforts to reduce anthropogenic greenhouse gases are needed to keep the amount of warming to levels deemed tolerable to society and ecosystems. Mitigation can come through changes in the actions of individuals, companies, and nations. Cost-effective actions, such as consuming a diet based on chicken and vegetables as opposed to a beef-based diet, can make as much of an impact on personal carbon emissions as switching from driving a standard to a hybrid vehicle (Stec and Cordero 2008).

Changes are also needed in the way we produce energy. In 2011, the IPCC stated that renewable energy could meet most of the world's energy supply by 2050 if it is supported by effective and efficient policies (IPCC 2011, 17–18, 21, 24–26). Demonstrating how to achieve this future, Germany is phasing out its nuclear energy facilities and shifting to renewable energy (Davidson 2012). Switzerland has stated it will phase out nuclear energy by 2034 (Mombelli 2012; swissinfo. ch 2013). Shifts away from nuclear are motivated by concerns over accidents, natural disasters, and the intractable problem of disposal of radioactive material. Nonetheless, some are calling for nuclear power as a major mitigator of emissions because the fission process itself does not produce greenhouse gases. Emissions of greenhouse gases, however, do result from the mining of fuels, building of facilities, and deconstructing of decommissioned nuclear power plants. Many remain adamantly opposed to nuclear energy. Energy efficiency, demand response, renewable energy (distributed and utility scale), storage, combined heat and power, and carbon capture and sequestration can decarbonize the energy sector without nuclear energy.

Greenhouse gas emissions can also be reduced through changes in agricultural processes. For example, sustainable biochar can be used as a soil amendment to improve soil quality and sequester carbon (International Biochar Initiative 2013a). A biochar protocol for the voluntary carbon-offset market is in review and is expected to be available for public comment after technical review is completed (International Biochar Initiative 2013b). Planting crops without tilling the soil, incorporating principles of conservation agriculture, is another method for reducing greenhouse gas emissions while improving crop resiliency (Food and Agriculture Organization of the United Nations 2010, 5–6).

Delayed enactment of mitigation forces us closer to temperature thresholds deemed detrimental to life on our planet. What happens when mitigation efforts fail, adaptation efforts are behind schedule, and severe consequences are imminent? Deliberately engineering the earth's climate system, or geoengineering, may be a last-ditch effort to counter warming. Thus far, geoengineering has primarily focused on methods to reduce the amount of solar radiation absorbed by the planet.

Osofsky and McAllister (2012) note that these methods generally fall into two categories: 1) solar radiation management, which involves increasing the planet's ability to reflect incoming sunlight; and 2) carbon dioxide removal, which involves the actual physical removal of carbon from the atmosphere in various ways. Solar radiation management measures work quickly to address the problem of warming, but do not address the underlying causes of the problem. Solar radiation management includes dispersing aerosols (or bright, reflective particles) into the stratosphere, putting into orbit a set of reflective solar shields, and deploying a fleet of unmanned spaceships to increase cloud formation in the subtropical oceans. Very danger-prone options, all. Increasing the reflectivity of the planet would theoretically reflect enough sunlight back to space to buffer anthropogenic warming. However, these methods would be costly and technologically difficult, would not necessarily prevent localized impacts, and could result in significant—and often unknown—safety and environmental consequences. For example, adding SO_2 particles to the atmosphere, as has been proposed, could further damage the ozone layer.

Carbon dioxide removal methods address the underlying causes of warming, but take effect much more slowly. The most commonly proposed measure includes the addition of micronutrients such as iron into the oceans to increase the capacity of phytoplankton to uptake carbon from seawater. As phytoplankton die, they sink to the seafloor, taking the excess CO_2 with them into the deep ocean and preventing the CO_2 from reentering the atmosphere. Iron is plentiful on earth, and the cost of such a carbon removal scheme would be relatively moderate. However, iron-fertilization schemes have thus far yielded disappointing results in removal of atmospheric carbon, and numerous uncertainties abound regarding maximizing the effectiveness of ocean-fertilization schemes. Furthermore, these efforts could result in substantial and unpredictable alterations to marine ecosystems. Because phytoplankton form the basis for marine food webs, changes to phytoplankton populations could result in repercussions throughout these food webs and may also cause increased levels of methane and other heat-trapping gases as phytoplankton conduct natural metabolic processes and as they decay after death (Osofsky and McAllister 2012).

In short, geoengineering is another global experiment, similar to the ongoing experiment to increase greenhouse gas concentrations. While experiments that go awry in a lab can be quickly tended to, global experiments are not quickly reversible, and pose the risk of potential unintended outcomes. Not only would geoengineering be costly, the potential consequences could amplify climate alterations and introduce new ones (e.g., Robock et al. 2009). Geoengineering could cause additional conflict at the international level. It could aggravate rather than alleviate climate-related stressors. And, in the process, it could divert sorely needed resources away from adaptation and mitigation efforts. The community of nations needs to consider if and when geoengineering should be enacted, who will pay for it, and, ultimately, who controls the dial.

Conclusion

Climate change was recognized as a global environmental problem in the late 1970s, spurring an international scientific effort to understand its causes and effects. Climate models have been developed that allow researchers to provide climate projections well into the future. These projections can then be translated into consequences for both human and natural systems. Some changes have already begun, while others will be pronounced toward the middle and end of the twenty-first century.

Effects of climate change are continually under study. Some uncertainty remains as to how much the planet will warm and the associated changes in climate and extreme events. Some uncertainty comes from specific knowledge gaps, such as about our understanding of the mechanisms that control the rate of ice loss in Greenland and Antarctica (so scientists may more accurately predict the range of future sea level rise). Refinements in these projections will result from scientific advances that will better inform adaptation or mitigation strategies. We have a problem with no easy solution. Although carbon is a building block for life on earth, a carbon dioxide molecule emitted into the atmosphere remains in the atmosphere for twenty to one hundred years. Changes have been set in motion. The rate of change is a crucial component in how our planet weathers the storm. Efforts to slow the rate of warming via mitigation may provide additional time for effective adaptation. But mitigation challenges are beyond the scope of a single individual, or even a single country. Still, individuals and individual countries can lead by example.

References

Abatzoglou, J. T. 2011. Influence of the PNA on declining mountain snowpack in the western United States. *International Journal of Climatology* 31:1099–1256. doi: 10.1002/joc.2137.

Abatzoglou, J. T., and C. A. Kolden. 2011. Climate change in western US deserts: Potential for increased wildfire and invasive annual grasses. *Rangeland Ecology and Management*. doi: 10.2111/REM-D-09-00151.1.

Abatzoglou, J. T., K. T. Redmond, and L. M. Edwards. 2009. Classification of regional climate variability in the state of California. *Journal of Applied Meteorology and Climatology* 48(8):1527–1541.

Ackerman, Frank. 2009. *Can We Afford the Future? The Economics of a Warming World*. New York: Zed Books.

Adger, Neil, Pramod Aggarwal, Shardul Agrawala, Joseph Alcamo, Abdelkader Allali, Oleg Anisimov, Nigel Arnell, et al. 2007. Climate change impacts, adaptation, and vulnerability: Summary for Policymakers. In *Climate Change 2007: Impacts, Adaptation and Vulnerability. Contribution of Working Group II to the Fourth Assessment Report of the Intergovernmental Panel on Climate Change*, edited by M. L. Parry, O. F. Canziani, J. P. Palutikof, P. J. van der Linden, and C. E. Hanson, 7–22. Cambridge, UK: Cambridge University Press.

Alley, Richard, Terje Berntsen, Nathaniel L. Bindoff, Zhenlin Chen, Amnat Chidthaisong, Pierre Friedlingstein, Jonathan M. Gregory, et al. 2007. IPCC 2007: Summary for Policymakers. In *Climate Change 2007: The Physical Science Basis. Contribution of Working Group I to the Fourth Assessment Report of the Intergovernmental Panel on Climate Change*, edited by S. Solomon, D. Qin, M. Manning, Z. Chen, M. Marquis, K. B. Averyt, M. Tignor, and H. L. Miller. New York: Cambridge University Press.

Amstrup, S. C., B. G. Marcot, and D. C. Douglas. 2007. *Forecasting the Rangewide Status of Polar Bears at Selected Times in the 21st century: Administrative Report*. Anchorage, AK: US Geological Survey, Alaska Science Center.

Anderson, J., F. Chung, M. Anderson, L. Brekke, D. Easton, M. Ejeta, R. Peterson, et al. 2008. Progress on incorporating climate change into management of California's water resources. *Climatic Change* 87 (Suppl. 1):S91–S108. doi: 10.1007/s10584-007-9353-1.

Archer, S. R., and K. I. Predick. 2008. Climate change and ecosystems of the southwestern United States. *Rangelands* 30(3):23–28. doi: 10.2111/1551-501X.

Baldocchi, D., and S. Wong. 2008. Accumulated winter chill is decreasing in the fruit growing regions of California. *Climatic Change* 87 (Suppl. 1):S153–S166. doi: 10.1007/s10584-007-9367-8.

Bograd, Steven J., Isaac Schroeder, Nandita Sarkar, Xuemei Qiu, William J. Sydeman, and Franklin B. Schwing. 2009. Phenology of coastal upwelling in the California Current. *Geophysical Research Letters* 36:L01602. http://130.207.67.194/web_db_papers/pobex_pdfs/Bograd-2009.pdf.

Borenstein, Seth. 2013. US scientists report big jump in heat-trapping CO_2. Associated Press, March 5. http://bigstory.ap.org/article/us-scientists-report-big-jump-in-heat-trapping-co2.

Brown, T. J., B. L. Hall, and A. L. Westerling. 2004. The impact of twenty-first century climate change on wildland fire danger in the western United States: An applications perspective. *Climatic Change* 62:365–388.

California Independent System Operator. 2013. Fast facts: What the duck curve tells us about managing a green grid. October. http://www.caiso.com/Documents/FlexibleResourcesHelpRenewables_FastFacts.pdf, accessed November 23, 2013.

Cayan, D. R., E. P. Maurer, M. D. Dettinger, M. Tyree, and K. Hayhoe. 2008. Climate change scenarios for the California region. *Climatic Change* 87 (Suppl. 1):S21–S42.

Cayan, Daniel R., Susanne Moser, Guido Franco, Michael Hanemann, and Myoung-Ae Jones. 2011. Second California assessment: Integrated climate change impacts assessment of natural and managed systems. *Climatic Change* 109 (Suppl. 1):S1–S19.

Chapin, F. S., III, M. Sturm, M. C. Serreze, J. P. McFadden, J. R. Key, A. H. Lloyd, A. D. McGuire, et al. 2005. Role of land-surface changes in Arctic summer warming. *Science* 310:657–660.

Christidis, N., G. C. Donaldson, and P. A. Stott. 2010. Causes for the recent changes in cold- and heat-related mortality in England and Wales. *Climatic Change* 102:539–553.

Climate Adaptation Knowledge Exchange. 2010. Relocating the Village of Newtok, Alaska due to Coastal Erosion. By Kirsten Feifel and Rachel M. Gregg. July 3. http://www.cakex.org/case-studies/1588. Accessed September 14, 2013.

Cloern, J. E., N. Knowles, L. R. Brown, D. Cayan, M. D. Dettinger, T. L. Morgan, D. H. Schoellhamer, et al. 2011. Projected evolution of California's San Francisco Bay-Delta-River system in a century of climate change. *PLoS ONE* 6(9):e24465. doi: 10.1371/journal.pone.0024465.

Cole, K., K. Ironside, J. Eischeid, G. Garfin, P. Duffy, and C. Toney. 2011. Past and ongoing shifts in Joshua tree support future modeled range contraction. *Ecological Applications* 21(1):137–149. e-View. doi: 10.1890/09-1800.

Cordero, E. C., W. Kessomkiat, J. T. Abatzoglou, and S. A. Mauget. 2011. The identification of distinct patterns in California temperature trends. *Climatic Change* 108:357–382. doi: 10.1007/s10584-011-0023-y.

Crimmins, S. M., S. Z. Dobrowski, J. A. Greenberg, J. Abatzoglou, and A. R. Mynsberge. 2011. Changes in climatic water balance drive downhill shifts in plant species' optimum elevations. *Science* 331:324–327.

Das, T., Dettinger, M., Cayan, D., and Hidalgo, H. 2011. Potential increase in floods in California's Sierra Nevada under future climate projections. *Climatic Change* 109 (Suppl. 1):S71–S94. doi: 10.1007/s10584-011-0298-z.

Davidson, Osha Gray. 2012. So far so good for Germany's nuclear phase-out, despite dire predictions. *Inside Climate News,* November 16. http://insideclimatenews.org/news/20121115/germany-energiewende-nuclear-energy-fukushima-chernobyl-merkel-renewables.

Davis, R. E., P. C. Knappenberger, W. M. Novicoff, and P. J. Michaels. 2003. Decadal changes in summer mortality in U.S. cities. *International Journal of Biometeorology* 47:166–175.

Diffenbaugh, N. S., and M. Ashfaq. 2010. Intensification of hot extremes in the United States. *Geophysical Research Letters* 37:L15701. doi: 10.1029/2010GL043888.

Diffenbaugh, N. S., M. A. White, G. V. Jones, and M. Ashfaq. 2011. Climate adaptation wedges: A case study of premium wine in the western United States. *Environmental Research Letters* 6:024024. doi: 10.1088/1748-9326/6/2/024024.

Durner, George M., David C. Douglas, Ryan M. Nielson, Steven C. Amstrup, Trent L. McDonald, Ian Stirling, Mette Mauritzen, et al. 2009. Predicting 21st-century polar bear habitat distribution from global climate models. *Ecological Monographs* 79:25–58. doi: 10.1890/07-2089.1.

Elsner, M. M., L. Cuo, N. Voisin, J. Deems, A. F. Hamlet, J. A. Vano, K. E. B. Mickelson, et al. 2010. Implications of 21st century climate change for the hydrology of Washington State. *Climatic Change* 102 (1–2):225–260. doi: 10.1007/s10584-010-9855-0.

Emanuel, Kerry. 2012. *What We Know About Climate Change.* 2nd ed. Cambridge, MA: MIT Press.

FEMA. 2013. New York recovery from Hurricane Sandy: By the numbers. Release Number: NR-210, April 19. http://www.fema.gov/news-release/2013/04/19/new-york-recovery-hurricane-sandy-numbers.

Field, C. B., G. C. Daily, F. W. Davis, S. Gaines, P. A. Matson, J. Melack, and N. L. Miller. 1999. *Confronting Climate Change in California: Ecological Impacts on the Golden State.* Washington, DC: Ecological Society of America.

Food and Agriculture Organization of the United Nations. 2010. Climate-smart agriculture: Policies, practices and financing for food security, adaptation, and mitigation. http://www.fao.org/docrep/013/i1881e/i1881e00.pdf.

Francis, Jennifer A., and Stephen J. Vavrus. 2012. Evidence linking Arctic amplification to extreme weather in mid-latitudes. *Geophysical Research Letters* 39:L06801.

Frazier, T., N. Wood, B. Yarnal, and D. Bauer. 2010. Influence of potential sea level rise on societal vulnerability to hurricane storm-surge hazards, Sarasota County, Florida. *Applied Geography (Sevenoaks, England)* 30:490–505. doi: 10.1016/j.apgeog.2010.05.005.

Gershunov, Alexander, Daniel R. Cayan, and Sam F. Iacobellis. 2009. The great 2006 heat wave over California and Nevada: Signal of an increasing trend. *Journal of Climate* 22:6181–6203. doi: 10.1175/2009JCLI2465.1.

Gershunov, A., and K. Guirguis. 2012. California heat waves in the present and future. *Geophysical Research Letters* 39:L18710. doi: 10.1029/2012GL052979.

Gillis, Justin. 2013a. Arctic Ice Makes Comeback From Record Low, but Long-Term Decline May Continue. *New York Times*. September 20. http://www.nytimes.com/2013/09/21/science/earth/arctic-ice-makes-come-back-from-record-low-but-long-term-decline-may-continue.html?_r=1&&pagewanted=print, accessed September 21, 2013.

Gillis, Justin. 2013b. How high could the tide go? *New York Times*. January 21. http://www.nytimes.com/2013/01/22/science/earth/see king-clues-about-sea-level-from-fossil-beaches.html?hp.

Goldenberg, Suzanne. 2013. America's first climate refugees. *The Guardian,* n.d. http://www.guardian.co.uk/environment/interactive/2013/may/13/newtok-alaska-climate-change-refugees, accessed June 27, 2013.

Gruber, Nicolas, Claudine Hauri, Zouhair Lachkar, Damian Loher, Thomas L. Frölicher, Gian-Kasper Plattner. 2012. *Rapid progression of ocean acidification in the California Current system. Science* 337:220–223.

Hamlet, Alan F., Se-Yeun Lee, Kristian E. B. Mickelson, and Marketa M. Elsner. 2010. *Climatic Change.* 102 (1–2):103–128.

Hayhoe, Katharine, D. Cayan, C. B. Field, P. C. Frumhoff, E. P. Maurer, N. L. Miller, S. C. Moser, et al. 2004. Emission pathways, climate change, and impacts on California. *Proceedings of the National Academy of Sciences* 101:12422–12427.

Intergovernmental Panel on Climate Change. 2007. *Climate Change 2007: The Physical Science Basis.* Contribution of Working Group I to the Fourth Assessment Report of the Intergovernmental Panel on Climate Change. Edited by S. Solomon, D. Qin, M. Manning, Z. Chen, M. Marquis, K. B. Averyt, M. Tignor, and H. L. Miller. Cambridge, UK and New York: Cambridge University Press.

Intergovernmental Panel on Climate Change. 2011. Summary for Policymakers. In IPCC Special Report on Renewable Energy Sources and Climate Change Mitigation, edited by O. Edenhofer, R. Pichs-Madruga, Y. Sokona, K. Seyboth, P. Matschoss, S. Kadner, T. Zwickel, et al. Cambridge, UK: Cambridge University Press. http://srren.ipcc-wg3.de/report/IPCC_SRREN_SPM.pdf.

Intergovernmental Panel on Climate Change. 2013. Working Group I Contribution to the IPCC Fifth Assessment Report Climate Change 2013: The Physical Science Basis Summary for Policymakers. http://www.climatechange2013.org/spm.

International Biochar Initiative. 2013a. http://www.biochar-international.org/sustainability.

International Biochar Initiative. 2013b. http://www.biochar-international.org/sites/default/files/March_2013_final.pdf.

Johnstone, J. A., and T. E. Dawson. 2010. Climatic context and ecological implications of summer fog decline in the coast redwood region. *Proceedings of the National Academy of Sciences of the United States of America* 107(10):4533–4538.

Kaplan, Thomas. 2013. State tells investors that climate change may hurt its finances. *New York Times*. March 26. http://www.nytimes.com/2013/03/27/nyregion/new-york-state-bonds-include-warning-on-climate-change.html?_r=1&&pagewanted=print, accessed June 27, 2013.

Karl, T. R., J. M. Melillo, and T. C. Peterson. 2009. *Global Climate Change Impacts in the United States.* Boston: Cambridge University Press.

Lafferty, Kevin D. 2009. The ecology of climate change and infectious diseases. *Ecology* 90:888–900. doi: 10.1890/08-0079.1.

Littell, J. S., E. E. Oneil, D. McKenzie, J. A. Hicke, J. Lutz, and R. A. Norheim. 2010. Forest ecosystems, disturbance, and climatic change in Washington State, USA. *Climatic Change* 102:129–158. doi: 10.1007/s10584-010-9858-x.

Loarie, S. R., P. B. Duffy, H. Hamilton, G. P. Asner, C. B. Field, and D. D. Ackerly. 2009. The velocity of climate change. *Nature* 462:1052–1105.

Lobell, D. B., M. B. Burke, C. Tebaldi, M. M. Mastrandrea, W. P. Falcon, and R. L. Naylor. 2008. Prioritizing climate change adaptation needs for food security in 2030. *Science* 319:607–610. doi: 10.1126/science.1152339.

Lutz, J. A., J. W. van Wagtendonk, and J. F. Franklin. 2010. Climatic water deficit, tree species ranges, and climate change in Yosemite National Park. *Journal of Biogeography* 37(5):936–950.

Medellin-Azuara, J., J. J. Harou, M. A. Olivares, K. Madani, J. R. Lund, R. E. Howitt, S. K. Tanaka, et al. 2008. Adaptability and adaptations of California's water supply system to dry climate warming. *Climatic Change* 87 (Supp. 1):S75–S90. doi: 10.1007/s10584-007-0355-z.

Meehl, G. A., J. M. Arblaster, and C. Tebaldi. 2005. Understanding future patterns of precipitation extremes in climate model simulations. *Geophysical Research Letters* 32:L18719.

Miller, N., K. Bashford, and E. Strem. 2003. Potential impacts of climate change on California. *Journal of the American Water Resources Association,* Hydrology Paper No. 02035. http://esd.lbl.gov/FILES/ about/staff/ normanmiller/miller_jawra2003.pdf.

Miller, Norman L., Katharine Hayhoe, Jiming Jin, and Maximilian Auffhammer. 2008. Climate, extreme heat, and electricity demand in California. *Journal of Applied Meteorology and Climatology* 47:1834–1844. doi: 10.1175/2007JAMC1480.1.

Mills, L. Scott, Marketa Zimovaa, Jared Oyler, Steven Running, John T. Abatzoglou, and Paul M. Lukacs. 2013. Camouflage mismatch in seasonal coat color due to decreased snow duration. *Proceedings of the National Academy of Sciences of the United States of America* 110(18):7360–7365. doi: 10.1073/pnas.1222724110.

Milly, P. C. D., Julio Betancourt, Malin Falkenmark, Robert M. Hirsch, Zbigniew W. Kundzewicz, Dennis P. Lettenmaier, Ronald J. Stouffer. 2008. Stationarity is dead: Whither water management? *Science* 319 (5863):573–574. doi: 10.1126/science.1151915.

Min, S.-K., X. Zhang, F. W. Zwiers, and G. C. Hegerl. 2011. Human contribution to more-intense precipitation extremes. *Nature* 470:378–381. doi: 10.1038/nature09763.

Mombelli, Armando. 2012. Major shift in energy policy looms. Swissinfo.ch, Swiss Broadcasting Corporation, November 15. http://www.swissinfo.ch/eng/swiss_news/Major_shift_in_energy_policy_looms. html?cid=33958136.

Mote, P. W., A. F. Hamlet, M. P. Clark, and D. P. Lettenmaier. 2005. Declining mountain snowpack in western North America. *Bulletin of the American Meteorological Society* 86:39–49.

Muller, Richard A. 2013. A pause, not an end, to warming. Op-Ed Contributor. Opinion Pages. *New York Times.* September 25. http://www.nytimes.com/2013/09/26/opinion/a-pause-not-an-end-to-warming. html?_r=0%20accessed%20September%2026,%20 2013&pagewanted=print, accessed September 27, 2013.

Munich Re Group. 2011. Weather extremes, climate change, Durban 2011. Electronic press folder. Status 25 November. http://www.munichre.com/app_pages/www/@res/pdf/media_relations/press_dossiers/durban_2011/press_folder_durban_2011_en.pdf, accessed September 22, 2013.

NASA. 2013. The current and future consequences of global change. http://climate.nasa.gov/effects/.

National Climate Assessment Development Advisory Committee. 2013. Draft Climate Assessment Report Released for Public Comment v. Jan 2013. http://ncadac.globalchange.gov/.

National Climatic Data Center. 2013. Billion-Dollar Weather/Climate Disasters. NOAA. http://www.ncdc. noaa.gov/billions/.

National Oceanic and Atmospheric Adminstration. 2012. http://oceanexplorer.noaa.gov/explorations/02quest/background/upwelling/upwelling.html.

National Research Council. 2010. Committee on the Development of an Integrated Science Strategy for Ocean Acidification Monitoring, Research, and Impacts Assessment. *Ocean Acidification: A National Strategy to Meet the Challenges of a Changing Ocean.* Washington, DC: The National Academies Press.

National Resources Defense Council. 2009. Ocean acidification: The other CO_2 problem. http://www.nrdc.org/oceans/acidification, accessed September 16, 2013.

National Resources Defense Council. 2013. Climate change health threats in New York. http://www.nrdc.org/health/climate/ny.asp#airpollution, accessed June 27, 2013.

New York State Department of Environmental Conservation. 2010. New York State Climate Action Plan Interim Report. http://www.dec.ny.gov/energy/80930.html, accessed March 29, 2013.

Ocean Carbon and Biogeochemistry Program, Ocean Acidification Subcommittee. 2012. FAQs about ocean acidification: Management and mitigation options. What can we do about ocean acidification on local and regional scales? September 24. http://www.whoi.edu/OCB-OA/page.do?pid=112161#5.

Ortiz, R., K. D. Sayre, B. Govaerts, R. Gupta, G. V. Subbarao, T. Ban, Hodson, D., et al. 2008. Climate change: Can wheat beat the heat? *Agriculture, Ecosystems & Environment* 126:46–58.

Osofsky, Hari M., and Lesley K. McAllister. 2012. *Climate Change Law and Policy.* New York: Wolters Kluwer.

Overland, James E., Jennifer A. Francis, Edward. Hanna, and Muyin. Wang. 2012. The recent shift in early summer Arctic atmospheric circulation. Atmospheric Science. *Geophysical Research Letters* 39(19), L19804, doi: 10.1029/2012GL053268.

Parris, A., P. V. Bromirski, D. Burkett, M. Cayan, J. Hall Culver, R. Horton, K. Knuuti, et al. 2012. Global sea level rise scenarios for the US national climate assessment. NOAA Tech Memo OAR CPO-1.

Rahmstorf, Stefan. 2013a. Sea-level rise: Where we stand at the start of 2013. Realclimate.org, January 9. http://www.realclimate.org/index.php/archives/2013/01/sea-level-rise-where-we-stand-at-the-start-of-2013.

Rahmstorf, Stefan. 2013b. Sea-level rise: Where we stand at the start of 2013. Part 2. Realclimate.org, January 9. http://www.realclimate.org/index.php/archives/2013/01/sea-level-rise-where-we-stand-at-the-start-of-2013-part-2/.

Rehfeldt, Gerald E.; Nicholas L. Crookston, Marcus V. Warwell, and Jeffrey S. Evans. 2006. Empirical analyses of plant-climate relationships for the western United States. *International Journal of Plant Sciences* 167(6):1123–1150.

Revkin, Andrew. 2013. Climate Panel's Fifth Report Clarifies Humanity's Choices. Dot Earth Blog. Opinion Pages. *New York Times.* September 2. http://dotearth.blogs.nytimes.com/2013/09/27/ipcc-global-warming-report-clarifies-humanitys-choices/?_r=1&&pagewanted=print.

Riordan, B., D. Verbyla, and A. D. McGuire. 2006. Shrinking ponds in subarctic Alaska based on 1950–2002 remotely sensed images. *Journal of Geophysical Research* 111:G04002. doi: 10.1029/2005JG000150.

Robock, A., A. Marquardt, B. Kravitz, and G. Stenchikov. 2009. Benefits, risks, and costs of stratospheric geoengineering. *Geophysical Research Letters* 36:L19703. doi: 10.1029/2009GL039209.

Rosenzweig, C., W. Solecki, A. DeGaetano, M. O'Grady, S. Hassol, and P. Grabhorn, eds. 2011. Responding to Climate Change in New York State: The ClimAID Integrated Assessment for Effective Climate Change Adaptation. Technical Report. Albany, NY: New York State Energy Research and Development Authority.

Roy, Sujoy B., Limin Chen, Evan Girvetz, Edwin P. Maurer, William B. Mills, and Thomas M. Grieb. 2010. Evaluating sustainability of projected water demands under future climate change scenarios. A Tetra Tech, Inc. Report. Prepared for the Natural Resources Defense Council. July. http://timmcgivern.files.wordpress.com/2010/07/tetra_tech_climate_report_2010_lowres.pdf, accessed September 22, 2013.

Rupp, D. E., P. W. Mote, N. Massey, C. J. Rye, R. Jones, and M. R. Allen. 2012. Did human influence on climate make the 2011 Texas drought more probable? *Bulletin of the American Meteorological Society* 93:1041–1067. doi: 10.1175/BAMS-D-11-00021.1.

Samenow, Jason. 2012. Study: Arctic ice loss may be making North America weather more extreme. Capital Weather Gang Blog. *The Washington Post.* Posted October 10. http://www.washingtonpost.com/blogs/capital-weather-gang/post/study-arctic-ice-loss-making-north-america-weather-more-extreme/2012/10/10/e2f79b88-1300-11e2-ba83-a7a396e6b2a7_blog.html, accessed September 21, 2013.

Samenow, Jason. 2013. Arctic warming and our extreme weather: no clear link new study finds. Capital Weather Gang Blog. *The Washington Post.* Posted August 19. http://www.washingtonpost.com/blogs/capital-weather-gang/wp/2013/08/19/arctic-warming-and-our-extreme-weather-no-clear-link-new-study-finds/, accessed September 21, 2013.

Sathaye, J. A., L. L. Dale, P. H. Larsen, G. A. Fitts, K. Koy, S. M. Lewis, and A. Frossard Pereira de Lucena. 2013. Estimating impacts of warming temperatures on California's electricity system. *Global Environmental Change* 23:499–511.

Schwartz, M. D., R. Ahas, and A. Aasa. 2006. Onset of spring starting earlier across the Northern Hemisphere. *Global Change Biology* 12:343–351. doi: 10.1111/j.1365-2486.2005.01097.x.

Seager, R., N. Pederson, Y. Kushnir, J. Nakamura, and S. Jurburg. 2012. The 1960s drought and the subsequent shift to a wetter climate in the Catskill Mountains region of the New York City watershed. *Journal of Climate*, 25:6721–6742.

Seager, R., M. Ting, I. Held, Y. Kushnin, J. Lu, G. Vecchi, H.-P. Huang, et al. 2007. Model projections of an imminent transition to a more arid climate in southwestern North America. *Science* 316:1181–1184.

Stec, L., and E. Cordero. 2008. *Cool Cuisine: Taking the Bite out of Global Warming.* Utah: Gibbs Smith.

Stott, P. A., D. A. Stone, and M. R. Allen. 2004. Human contribution to the European heatwave of 2003. *Nature* 432:610–614.

swissinfo.ch. 2013. Energy change is already underway, says Leuthard. September 4. http://www.swissinfo.ch/eng/science_technology/Energy_change_is_already_underway,_says_Leuthard.html?view=print&cid=36822514, accessed September 22, 2013.

United Nations Environmental Programme. 2012. 14th Global Meeting of the Regional Seas Conventions and Action Plans Nairobi, Kenya, 1st–3rd October 2012. http://www.unep.org/regionalseas/global meetings/14/RS.14_WP.4.RS.pdf

Warner, Robin, and Clive Schofield, eds. 2012. *Climate Change and the Oceans: Gauging the Legal and Policy Currents in the Asia Pacific and Beyond.* Northampton, MA: Edward Elgar.

Webster, P. J., G. J. Holland, J. A. Curry, and H. R. Chang. 2005. Changes in tropical cyclone number, duration, and intensity in a warming environment. *Science* 309:1844–1846.

Weiss, Kenneth R. 2012. Oceans' rising acidity a threat to shellfish—and humans. *Los Angeles Times.* October 6.

Westerling, A. L., B. P. Bryant, H. K. Preisler, T. P. Holmes, H. Hidalgo, T. Das, and S. Shrestha. 2011a. Climate change and growth scenarios for California wildfire. *Climatic Change* 109(s1):445–463.

Westerling, A. L., M. G. Turner, E. H. Smithwick, W. H. Romme, and M. G. Ryan. 2011b. Continued warming could transform Greater Yellowstone fire regimes by mid-21st Century. *Proceedings of the National Academy of Sciences of the United States of America* 108(32):13165–13170.

Wetherald, Richard T., and Syukuro Manabe. 2002. Simulation of hydrologic changes associated with global warming. *Journal of Geophysical Research* 107:4379.

White, M. A., N. S. Diffenbaugh, G. V. Jones, J. S. Pal, and F. Giorgi. 2006. Extreme heat reduces and shifts United States premium wine production in the 21st century. *Proceedings of the National Academy of Sciences of the United States of America* 103:11217–11222.

Wilkinson, R., K. Clarke, M. Goodchild, J. Reichman, and J. Dozier. 2002. *The Potential Consequences of Climate Variability and Change for California: The California Regional Assessment.* Washington, DC: US Global Change Research Program.

Williams, A. P., C. D. Allen, C. Millar, T. Swetnam, J. Michaelsen, C. J. Still, and S. W. Leavitt. 2010. Forest responses to increasing aridity and warmth in southwestern North America. *Proceedings of the National Academy of Sciences of the United States of America* 107(50):21289–21294.

Williams, J. W., S. T. Jackson, and J. E. Kutzbacht. 2007. Projected distributions of novel and disappearing climates by 2100 AD. *Proceedings of the National Academy of Sciences of the United States of America* 104:5738–5742.

Wootton, J. Timothy, Catherine A. Pfister, and James D. Forester. 2008. Dynamic patterns and ecological impacts of declining ocean pH in a high-resolution multi-year dataset. *Proceedings of the National Academy of Sciences of the United States of America* 105(48): 18848–18853.

Discussion Questions

1   Climate disruption is regarded as an additional stressor that challenges human populations that are already struggling. In what ways has this already occurred, and in what ways is it likely to occur? That is, how do more extreme climate conditions, including extreme weather events, impact human well-being and needs?

2   Plants and animals are already documented as having changed where they live, i.e., their range, due to climate change, and more of this change will occur. What are some examples? Why should we be concerned?

3   Although it may seem contrary to a warming planet, why will some places, like western New York state, sometimes receive more snow with warmer winters?

4   Why are the impacts of three hurricanes in New York during the previous decade helpful in planning for continued climate disruption?

PART III

SPECIES UNDER THREAT IN THE CLIMATE CRISIS

CHAPTER 7

Madrean Sky Islands, North America

Laura López-Hoffman and Adrian Quijada-Mascareñas

The steep mountains, sheltered canyons, and grassland valleys of the Madrean Sky Island region of southwest United States and northwest Mexico is one of the world's most biologically diverse, temperate forest regions. A two-century legacy of mining and ranching has entailed significant impacts on the region's biodiversity, and today alterations due to climate change are already having visible impacts on species, ecosystems, and ecological connectivity. These changes, coupled with the global change drivers of increased urbanization and border security activities, will necessitate adaptive and forward-thinking strategies to conserve the region's biodiversity and to maintain connectivity across the border. The social, economic, and political challenges of straddling a border between two very different countries spotlight the need to foster both the institutional and scientific capacity for developing transborder conservation strategies that are adaptive to climate change.

Introduction to the Region

The term *sky islands* is widely used to describe continental mountain complexes, or inland archipelagos, comprising isolated mountain top "islands" surrounded by lower elevation valley "seas." The Madrean Sky Islands consist of roughly forty disjunct mountain complexes covering 180,000 square kilometers (Figure 7.1), with mountaintop elevations ranging from 1,800 to 3,267 meters. The mountains are separated by 15 to 25 kilometer-wide valleys of semidesert grassland, Sonoran desert upland, and Chihuahuan desert scrub. Valley elevations range from 750 to 1,400 meters, increasing west-to-east toward the Continental Divide.

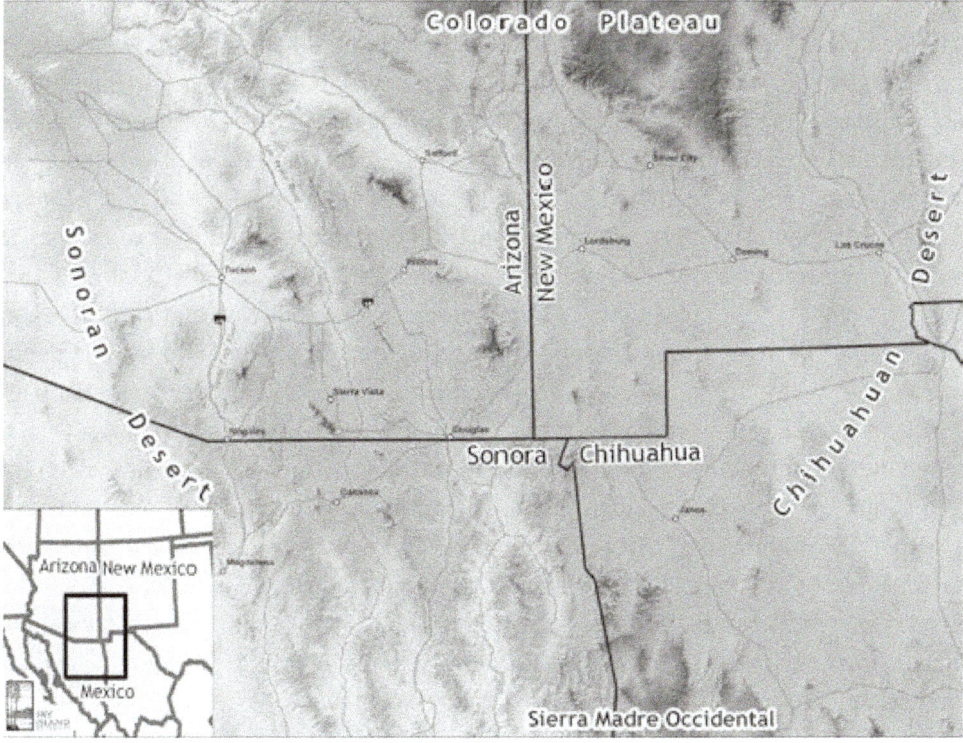

Figure 7.1 Map of Sky Island region (www.skyislandalliance.org).

The high degree of biodiversity in the Sky Islands region is due to two key factors: (1) steep elevation gradients and complex topography; and (2) biogeographic confluences between two climatic regions (temperate and subtropical), between two deserts (the Sonoran and Chihuahuan), and between two major cordilleras (the Rocky Mountains and Sierra Madre Occidental) (Riemann and Ezcurra 2007). Within the 250 kilometers span between the Rocky Mountains and Sierra Madre Occidental, many temperate and tropical species overlap at the edges of their distribution. The Sky Islands are the northern boundary for fourteen frost-susceptible subtropical plant family species (Felger and Wilson 1994), and eleven bird families reach either their northern or southern limits in the area (Warshall 1994).

Historical Overview of Conservation and Science Initiatives

The Sky Island landscape of today is the result of centuries, if not millennia, of interaction between people and the environment. Evidence of human activities dating back 10,000 years can be found from mountaintops to valley floors. Early inhabitants were likely seminomadic people who gathered plant resources and hunted big game, including mammoths. There is evidence of farming and substantial irrigation projects in the region by 1000 AD (Spoerl and Ravesloot 1994).

Ranching, one of the main drivers of environmental change in the region, originated with the Spanish colonial settlements and missions that began in the mid-1600s. The Spanish settlers raised livestock and spread them widely, clearing land for cattle grazing and raising wheat. The colonial period lasted until 1821, when the region became part of the new country of Mexico following its independence from Spain. Following the Treaty of Guadalupe Hidalgo in 1848 and the Gadsden Purchase in 1853, a newly drawn border divided the two countries, although both focused on exploiting the region's untapped riches. Irrigated agriculture and ranching expanded in both countries, and landscape fragmentation on both sides of the border accelerated as large tracks of land were cleared for cattle grazing. Mining, heretofore conducted on a small scale, burgeoned into an important regional industry, resulting both in the disappearance of hills and mountains into the expanding copper mine pits, and in the appearance of new mining boom-towns, such as Tombstone and Bisbee in Arizona and Cananea in Sonora. By the beginning of the 1900s, cattle barons and mine owners ruled the region (Varady and Ward 2009).

The land reform that followed the Mexican Revolution (~1910–20) entailed critical environmental consequences for the region: large ranches—many dating to Spanish colonial land grants—were partitioned into smaller, communally held lands called *ejidos*. As a result, the landscape was further fragmented and natural resources diminished by subsistence hunting, overgrazing of livestock, and timber exploitation (Culver et al. 2009). The environmental legacy of mining and livestock, including deforestation around mines, soil erosion in overgrazed areas, and diminished species populations, today remains a challenge to maintaining the region's ecological connectivity.

Between 1902 and 1907, forest reserves were declared throughout most of the Sky Islands of the United States, eventually becoming integrated into the Coronado National Forest. In 1936, Mexico designated the Sierra Los Ajos, Buenos Aires y La Púrica National Forest Reserve, and then in 1939 the Bavispe National Forest and Wildlife Reserve, the three separate areas of which are today collectively referred to as the Ajos-Bavispe Reserve (Búrquez & Martínez-Yrízar 2006). The reserve encompassed eight Sky Islands and the three most important watersheds in the state of Sonora: the Sonora, Bavispe-Yaqui, and San Pedro river basins. During the early part of the twentieth century, reserves in both countries were used to support timber and livestock production; as a result, most old-growth pine forests in the region disappeared by midcentury.

In 1994, in one of the first applied manifestations of the new "cores and corridors" approach to landscape conservation, conservationist Ton Povolitis designed a reserve plan for the US portion of the region (Povilitis 1994). The Sky Island Alliance and Wildlands Project followed in 2000 with a comprehensive framework for binational biodiversity conservation based on the protection of additional wilderness lands, the reintroduction of native species, and the conservation of landscape linkages between Sky Island ranges (Foreman et al. 2000). In 2004, The Nature Conservancy released an alternative conservation blueprint that highlighted priorities in valley locations centered on remaining grasslands and cienegas (Marshall et al. 2004). This plan in part led to the purchase of Rancho los Fresnos in northern Sonora, which is owned and comanaged by Naturalia, a Mexican nongovernmental organization (NGO) working to restore populations of the Mexican gray wolf and jaguar. In the same year, Conservation International added the Madrean Pine-Oak Woodlands region, of which the Sky Island region forms the northernmost portion, to its global list of Biodiversity Hotspots (Mittermeier et al. 2004).

Current Conservation in the Region

In the United States portion of the Sky Island region, the majority of higher elevation areas (>4,000 feet) are managed by the 1.78 million-acre Coronado National Forest, including ten wilderness areas. The US National Park Service also manages large wilderness areas within Saguaro National Park and Chiricahua National Monument. In all, 595,000 acres are managed as wilderness across different US federal jurisdictions. In Mexico, the main protected area is the Ajos-Bavispe Reserve, encompassing 445,000 acres (1,800 square kilometers) across three areas in northern Sonora. Due to its ruggedness, the Ajos-Bavispe Reserve has remained relatively protected.

In both countries, Sky Island valleys contain a mosaic of land tenures. In the United States, this includes lands held by the Bureau of Land Management, state trusts, and private owners. Within Mexico, most lands outside of the Ajos-Bavispe Reserve are privately or communally owned. Across all jurisdictions and in both countries, including privately owned land, approximately 1,500,000 acres of land are managed for conservation, representing 10 percent of the total land base.

Mining, exurbanization, and ranching remain forces of landscape change in the Sky Islands today. Mine development and expansion continues to grow due to increasing worldwide demand for copper, leading to controversy in both countries about environmental impacts (Hartman and Farr 2010; Reuters 2011). In Arizona, rapid development and exurban growth constrains conservation options for maintaining wildlife linkages between Sky Island mountain ranges. The US portion of the region is currently growing in human population at ~25 percent per decade, roughly two and a half times the national average. The growth is partly due to retirees from other parts of the country who are drawn to the region's moderate winter climate. In Mexico, growth is concentrated within existing urban areas and is not as much a threat to landscape connectivity in the Sky Islands.

Although ranching has degraded land in the region, many conservationists are examining ranching in a new light, championing the idea of "working landscapes." This growing perspective is based on the understanding that well-stewarded ranches can act as ecological buffers between protected areas and the pressures of urban areas (Knight 2009), and that the accelerated subdivision of ranch lands for exurban residential and recreation development will cause more serious and long-lasting habitat fragmentation than careful ranching (Nabhan et al., forthcoming). In the United States, a prominent example of this new approach is the Malpai Borderlands Group, a landowner-driven organization promoting ecosystem management for unfragmented landscapes in southeastern Arizona and southwestern New Mexico. In the Mexican portion of the region, a growing number of landowners, such as Carlos Robles of El Aribabi ranch in Sonora, are managing their ranches for small game hunting and bird watching, as well as cattle raising. In March 2011, the Mexican National Commission of Natural Protected Areas (CONANP) designated El Aribabi ranch as a Natural Protected Area under the Voluntary Land Conservation program. The designation protects 10,000 acres for biodiversity conservation, environmental education, and ecotourism. Such examples offer promise of ways to maintain ranching, ecological connectivity, and biodiversity in light of other threats.

The challenge of maintaining landscape and wildlife habitat permeability across the United States–Mexico border has been recently complicated by the surge in border security infrastructure and activities in the last decade (Flesch et al. 2010; Quijada-Mascareñas et al. 2011). Since 2000, pedestrian fencing (solid walls) and Normandy-style (i.e., antitank) vehicle barriers have been constructed on more than 60 percent of the United States–Mexico border. The barriers are accompanied by access roads—essentially swaths of cleared land up to sixty feet wide (Laura Lopez-Hoffman, pers. observ.). In the Sky Island region, the pedestrian fencing is mostly restricted in urban areas, with vehicle barriers in rural areas. As most areas of rugged, steep terrain do not have barriers, wildlife habitat connectivity has been mostly impaired in valleys. The lack of connectivity is particularly of concern for species such as black bear, jaguar, ocelot, and bob cats, which are more abundant on one side of the border than the other. This imbalance would put the less-abundant population at risk of extirpation if connectivity were cut off across the border (Culver et al. 2009). Because the Real ID Act of 2005 gave the secretary of the US Department of Homeland Security the right to waive laws in order to hasten border wall and road construction, environmental impact assessments on the impacts of wildlife and habitat have not been undertaken, as would normally have been the case under the US National Environmental Policy Act. An effort led by scientists from the US Geological Survey to develop a comprehensive protocol for monitoring the environmental impacts of border infrastructure and activities on wildlife, vegetation, and transboundary watershed hydrology has been languishing in review.

Regional Effects of Climate Change

The Sky Islands' climate is generally arid and warm, punctuated by rain arriving from the west in winter and cyclonic storms arriving from the south in summer (Adams and Comrie 1997; Sheppard et al. 2002). Mean annual temperatures and precipitation levels vary with elevation and range from 8 to 23 degrees Celsius and 250 to 800 millimeters, respectively. The percentage of total precipitation falling in winter decreases from west to east, whereas the percentage falling in summer decreases from south to north.

The Intergovernmental Panel on Climate Change and regional analyses suggest that the area encompassed by southwestern United States and central and western Mexico is likely to undergo significant precipitation and temperature changes throughout the twenty-first century. Mean annual precipitation is projected to decrease 5 to 10 percent, with the greatest decrease during winter months (Solomon et al. 2009; Overpeck and Udall 2010). Mean annual temperatures are increasing and projected to continue through coming decades, with consensus of a 3.0 to 3.5 degree Celsius annual mean temperature increase by 2099. This increase will not be evenly distributed throughout the year; summer high temperatures are projected to increase more than summer average temperatures, indicating greater variability in temperature extremes (IPCC-WG1 2007; Overpeck and Udall 2010).

Changing precipitation and temperature regimes are expected to affect the ecology of the Sky Islands in two main ways: (1) in the near term, fire regimes are expected to change, facilitating the spread of invasive species and eventual homogenization of the species pool; (2) over time, there will be shifts in species bioclimatic habitat envelopes—geographic spaces of climatically suitable habitat—which, within the physiography of Sky Islands, may result in "winning" species or "losing" species (i.e., local extirpations).

Fire is a fundamental ecological-organizing process in the Sky Islands. Grassland and forest fires regulate the distribution and abundance of vegetation, cycling of water and nutrients, soil formation and erosion, and carbon dynamics (DeBano and Ffolliott 1995; Falk et al. 2007). Fires influence ecosystem-level response to changing environments by resetting successional processes (Suding and Goldberg 2001). A changing climate may alter weather variables of importance to fire. For example, current projections suggest substantially decreased snowpack depth, density, and persistence, causing earlier soil drying and potentially longer fire seasons. Changes in the timing and total rainfall of the summer monsoon could also influence the prevalence of thunderstorms and, in turn, lightning ignitions of fires (Lin et al. 2007). Climate change may also alter the amplitude and duration of drought episodes: if mean temperatures increase, then the probability of extended periods above temperature and flammability thresholds will also increase (Brown et al. 2004).

The abundance of invasive, nonnative grasses at low elevations, such as buffel-grass (*Pennisetum ciliare*), may create new fire-spread pathways into higher elevation areas. As previously low-elevation fire regimes become more frequent at higher elevations, new pathways will be created for the spread of invasive species into higher elevations and the species pool of higher elevation areas might become homogenized as invasive grasses become more prevalent.

Along montane elevation gradients, area increases with ascent from the lowest elevations until reaching a centroid point (area-weighted midpoint). Above the centroid, available area decreases with ascent (Quijada-Mascareñas et al., in review). Due to this relationship between area and elevation, some Sky Island species will experience an expansion in habitat area with temperature-driven ascent under warming climate, whereas some species will experience a decrease in habitat area on a given mountain. Under ideal conditions, the latter would be able to shift to higher latitude mountain ranges with the same bioclimatic envelope conditions (Parmesan et al. 1999). However, the ability to shift latitudinally requires the ability to disperse across valleys beset by habitat loss and fragmentation due to urban growth, agriculture, mining, and ranching. Thomas et al. (2004) suggest that montane, range-restricted species that are impeded from free dispersal are at risk of extinction.

Approaches to Conservation under Climate Change

In a region such as the Sky Islands, which is divided by an international border, efforts to maintain landscape connectivity in the face of drivers of global change—climate change, exurbanization, border security, and mining—must transcend the political line. Despite the difficulties involved in developing cross-border conservation (Chester 2006; López-Hoffman et al. 2009), recent scholarship on climate change and water issues across the United States–Mexico border suggests that regional and local efforts, spearheaded by collaborations between civil society and governmental rather than purely national diplomatic initiatives, can build cross-border resilience to climate change (Wilder et al. 2010).

Three factors are critical to effective cross-border adaptive capacity (Wilder et al. 2010):

- Shared social learning is a common understanding of challenges among individuals and/or institutions. In this context, "platforms for shared social learning" refers to mechanisms by which the development of common conceptual understandings of climate change challenges can occur from the scale of individuals and NGOs to that of state and regional authorities across the international border.

- "Communities of practice" are bridges of information flow (i.e., networks) both across the border and across existing communities (i.e., associations based on shared identity, values, and practices). They are facilitated by individuals described as "network weavers" who can bring communities from both countries together (Laird-Benner and Ingram 2011), or by "boundary objects," meaning meetings or documents joining distinct communities in a common purpose (Wenger 1999).

- "Coproduction of knowledge" arises from a synergistic relationship between researchers on the one hand and stakeholders on the other to create "usable" science for framing policy (Lemos and Morehouse 2005). In a cross-border context, coproduction of knowledge requires decision makers, stakeholders, and scientists from both countries to work together to produce policy-relevant research.

Grounded in these three factors, the following discussion briefly reviews initiatives within each country to address landscape-level impacts of global change (in particular climate-related changes), and then provides a preliminary assessment of the present adaptive capacity of regional conservation initiatives and institutions. It concludes with recommendations for enhancing this capacity in the face of future changes.

Several recent initiatives by the US Department of the Interior (DOI) promise to support the development of adaptive capacity within the US portion of the Sky Island region. Under secretarial order 3289, the DOI mandated the establishment of eight Climate Science Centers around the country to promote climate change research. One of the centers will be established at the University of Arizona's School of Natural Resources and Environment, making it likely that new research will focus on the Sky Island region (DOI 2011). At the time of this writing, given that the center has yet to form, it is not clear whether the center's research will focus on basic science on climate change in the region or will engage in "coproducing" interdisciplinary knowledge that is directly applicable to policy making.

The order also mandated the establishment of Landscape Conservation Cooperatives (LCCs). The cooperatives are the applied science side of the DOI's plan to develop a coordinated, science-based response to climate change impacts on land, water, and wildlife resources. Cooperatives are intended to be communities of practice—termed "management-science" partnerships by the agency—involving scientists from academia, agencies, and NGOs that promise to support shared social learning about climate change (USFWS 2011). The Bureau of Reclamation and the US Fish and Wildlife Service are currently in the process of defining the Desert LCC, which is to encompass portions of five US states, three deserts (Mojave, Sonoran, and Chihuahuan), and several large river systems, including the Colorado River Basin. The agency would like to eventually consider portions of the Mexican states of Sonora, Chihuahua, and Coahuila as well. Some of the stated goals of the Desert LCC are to understand the effects of long-term drought on the composition, abundance, and distribution of species; reduced water availability on vegetation, wildlife, and human populations; increased temperatures on insect outbreaks and tree mortality; soil dryness and increasing air temperature on wildfire susceptibility; and changing fire regimes caused by increased invasion of nonnative grasses. As of spring 2011, the LCC staff has held several scoping meetings with agency, conservation organizations, and university stakeholders and has formed a stakeholder-steering committee (BOR 2011). If developed as intended, the Desert LCC should create a platform for shared social learning and a community of practice around climate change adaptation within the United States. While an agency official from Mexico's Instituto Nacional de Ecologica is on the steering committee, it remains to be seen how effectively Mexican stakeholders will be incorporated.

In addition to the Department of the Interior initiatives, which are still in a formative stage, the Sky Island Alliance, a regional NGO, is taking a lead in facilitating a community of practice on adaptation to climate change. Specifically, this group is in the process of a four-year initiative to identify organizational and landscape vulnerabilities to climate change and to develop strategies for addressing vulnerabilities. It is convening a series of workshops that are serving as boundary objects for bringing together diverse stakeholders from other NGOs, agency officials, and university researchers. The group held its first workshop in late 2010 and intends to hold two more workshops over the next two years. While most of the participants to date have been from the United States, the Sky Island Alliance organizers are committed to developing a cross-border dialogue in future workshops. Several of the organization's staff and board members (e.g., Sergio Avila and one of the coauthors, A. Quijada) have deep ties with Mexican NGOs, stakeholders, and agencies, and they should be able to function as cross-border network weavers.

On the Mexican side of the border, progress is being made to develop strategies for adapting to climate change (SEMARNAT-CONANP 2010; Locatelli et al. 2011). In accordance with the strategic objectives of Mexico's National Program for Protected Areas 2007–12, the National Commission for Protected Areas has developed a Climate Change Strategy for Protected Areas. The objectives of this program include (1) increasing the adaptive capacity of ecosystems and the communities in protected areas in the face of climate change, and (2) contributing to the mitigation of greenhouse gases and enhancement of carbon stocks via carbon-capture strategies. While the plan establishes strategies and guidelines for the commission's management decisions, it recognizes the importance of incorporating key stakeholders in such processes as well as strengthening technical and institutional capacities in climate change issues. The plan will be initially implemented in the Sierra Madre Sky Islands but has the potential to be applied toward other binational efforts in the region in the coming years.

Conclusion and Recommendations

As human activities increasingly fragment the landscape and reorganize species composition and distribution in the Sky Islands, concerted action is needed to develop binational strategies for dealing with (1) changing fire regimes, (2) invasive species control, and (3) the genetic isolation and potential loss of species due to constricted ranges as a result of both climate change and border infrastructure. Government agencies, conservation leaders, and organizations in the Sky Islands are making great progress in developing communities of practice and platforms for shared learning about environmental and climate-related drivers of change within each country. A next step in these efforts is to develop mechanisms to coproduce usable scientific knowledge for framing policy. Furthermore, these organizations and initiatives must begin to create the cross-border collaborations necessary to build binational adaptive capacity and maintain future ecological connectivity across the border.

Ultimately, it is critical that the Sky Island Alliance and the Desert Landscape Conservation Cooperative live up to their stated intentions of developing the cross-border communities of practice and platforms for shared social learning that will be fundamental in devising effective binational strategies for protecting transboundary connectivity. In this regard, it is incumbent upon the individuals within those organizations who have binational expertise to rise to the challenge and become border weavers, capable of fashioning cross-border, collaborative strategies for dealing with the global-change drivers—ranching, mining, border security, and climate change—that threaten the region.

References

Adams, D. K. and A. C. Comrie. 1997. "The North American Monsoon." *Bulletin of the American Meteorological Society* 78:2197–2213.

BOR (Bureau of Reclamation). 2011. *Landscape Conservation Cooperatives (LCCs): The Desert and Southern Rockies (LCC)*. Washington, DC: US BOR.

Brown, T. J., B. L. Hall, and A. L. Westerling. 2004. "The Impact of Twenty-first Century Climate Change on Wildland Fire Danger in the Western United States: An Applications Perspective." *Climatic Change* 62:365–388.

Búrquez, A. and A. Martínez-Yrízar. 2006. "Conservation and Landscape Transformation in Sonora, Mexico." In *Dry Borders: Great Natural Reserves of the Sonoran Desert*, R. S. Felger and B. Broyles, eds. Salt Lake City: University of Utah Press, 537–47.

Chester, C. C. 2006. *Conservation across Borders: Biodiversity in an Interdependent World*. Washington, DC: Island Press.

Culver, M., C. Varas, P. M. Harveson, B. McKinney and L. A. Harveson. 2009. "Connecting Wildlife Habitats across the US–Mexico Border." In *Conservation of Shared Environments: Learning from the United States and Mexico*, L. López-Hoffman, E. McGovern, R. G. Varady, and K. W. Flessa, eds. Tucson: University of Arizona Press, 83–100.

DeBano, L. F. and P. Ffolliott. 1995. *Biodiversity and Management of the Madrean Archipelago: The Sky Islands of Southwestern United States and Northwestern Mexico*. Fort Collins, CO: USDA Forest Service, Rocky Mountain Research Station, RM-GTR-264.

DOI (Department of the Interior). 2011. USDOI. *Climate Science Centers*. www.fws.gov/southeast/LCC/GulfPlains/pdf/GCPO%20Workshop_Jones%20&%20Shipp_Climate%20Science%20and%20LCCs_Mar10.pdf (accessed January 11, 2012).

Falk, D. A., C. Miller, D. McKenzie, and A. E. Black. 2007. "Cross-scale Analysis of Fire Regimes." *Ecosystems* 10:809–26.

Felger, R. S. and M. F. Wilson. 1994. "Northern Sierra Madre Occidental and Its Apachian Outliers: A Neglected Center of Biodiversity." In *The Sky Islands of Southwestern United States and Northwestern Mexico*, L. F. Debano, G. J. Gottfried, R. H. Hamre, C. B. Edminster, P. F. Ffolliott, and A. Ortega-Rubio, eds. USDA Forest Service: General Technical Report RM-GTR-264, 6–18.

Flesch, A. D., C. W. Epps, J. W. Cain, M. Clark, P. R. Krausman, and J. R. Morgart. 2010. "Potential Effects of the United States–Mexico Border Fence on Wildlife." *Conservation Biology* 24:171–81.

Foreman, D., B. Dugelby, J. Humphrey, B. Howard, and A. Holdsworth. 2000. "The Elements of a Wildlands Network Conservation Plan." *Wild Earth* 10:17–30.

Hartman, G. and M. Farr. 2010. "Rosement Mine Benefits Are Small and Short-term, Negatives Are Lasting and Costly." *Arizona Daily Star*. March 9.

IPCC-WG1. 2007. *Climate Change 2007: The Physical Science Basis*. Contribution of Working Group I to the fourth assessment report of the IPCC. New York: Cambridge University Press.

Knight, R. 2009. "The Wisdom of the Sierra Madre: Aldo Leopold, the Apaches and the Land Ethic." In *Conservation of Shared Environments: Learning from the United States and Mexico*, L. López-Hoffman, E. McGovern, R. G. Varady, and K. W. Flessa, eds. Tucson: University of Arizona Press, 71–77.

Laird-Benner, W. and H. Ingram. 2011. "Sonoran Desert Network Weavers." *Environment Magazine* 53:7–16.

Lemos, M. C. and B. J. Morehouse. 2005. "The Co-production of Science and Policy in Integrated Climate Assessments." *Global Environmental Change* 15:57–68.

Lin, J. L., E. Mapes, K. M. Weickmann, G. N. Kiladis, S. D. Schubert, M. J. Suarez, J. T. Bacmeister, and M. Lee. 2007. "North American Monsoon and Convectively Coupled Equatorial Waves Simulated by IPCC AR4 Coupled GCMs." *Journal of Climate* 21: 2919–37.

Locatelli B., V. Evans, A. Wardell, A. Andrade, and R. Vignola. 2011. "Forests and Climate Change in Latin America: Linking Adaptation and Mitigation." *Forests* 2:431–50.

López-Hoffman, L., E. D. McGovern, R. G. Varady, and K. W. Flessa. 2009. *Conservation of Shared Environments: Learning from the United States and Mexico.* First volume in a new series, *Edge: Environmental Science, Law and Policy*, M. Miller, B. Morehouse and J. Overpeck, eds. Tucson: University of Arizona Press.

Marshall, R. M., D. Turner, A. Gondor, D. Gori, C. Enquist, G. Luna, and R. Paredes Aguilar. 2004. *An Ecoregional Analysis of Conservation Priorities in the Apache Highlands Ecoregion.* Prepared by The Nature Conservancy of Arizona, Instituto del Medio Ambiente y el Desarrollo Sustentable del Estado de Sonora.

Mittermeier, R. A., P. R. Gil, M. Hoffman, J. Pilgrim, T. Brooks, C. G. Mittermeier, J. Lamoreux, and G. A. B. da Fonseca. 2004. *Hotspots Revisited.* Mexico City: CEMEX.

Nabhan, G. P., L. López-Hoffman, C. K. Presnall, R. Knight, J. Goldstein, H. Gosnell, L. Gwen, D. Thilmany, and S. Charnley. Forthcoming. "Payments for Ecosystem Services: Keeping Working Lands in Working Hands." In *Saving the Wide-Open Spaces*, T. Sheridan, S. Charnley, and G. P. Nabhan, eds. Tucson: University of Arizona Press.

Overpeck, J. and B. Udall. 2010. "Dry Times Ahead." *Science* 328:1642–43.

Parmesan, C., N. Ryrholm, C. Stefanescu, J. K. Hill, C. D. Thomas, H. Descimon, and B. Huntley, et al. 1999. "Poleward Shift of Butterfly Species' Ranges Associated with Regional Warming." *Nature* 399:579–83.

Povilitis, T. 1994. "A Nature Reserve System for the Gila River–Sky Island Region of Arizona and New Mexico; Some Preliminary Suggestions." In The *Sky Islands of Southwestern United States and Northwestern Mexico*, L. F. Debano, G. J. Gottfried, R. H. Hamre, C. B. Edminster, P. F. Ffolliott, and A. Ortega-Rubio, eds. USDA Forest Service: General Technical Report RM-GTR-264, 6–18.

Quijada-Mascareñas, A., D. Falk, J. Weiss, M. McClaran, J. Koprowski, M. Culver, S. Drake, S. Marsh, W. van Leeuwen, and M. Skroch. In review. "Potential Multi-scale Ecological Effects of Climate Change in Sky Islands." *Climate Research.*

Quijada-Mascareñas, J. A., C. Van Riper, D. James, L. López-Hoffman, C. Sharp, R. Gimblett, M. Scott, L. Norman, et al. 2011. "An Ecosystem Approach for Monitoring the Border Wall." In *The Border Wall between Mexico and the United States. Venues, Mechanisms and Stakeholders for a Constructive Dialogue*, C. de la Parra, E. Peters, and A. Cordova y Vazquez, eds. Instituto Nacional de Ecología, Mexico: SEMARNAT, Mexico DF.

Reuters. 2011. *Correction-Update-Grupo Mexico: Cananea at Capacity in March.* 8 February. www.reuters.com/article/2011/02/08/grupomexico-idUSN0829272520110208 (accessed January 10, 2012).

Riemann, H. and E. Ezcurra. 2007. "Endemic Regions of the Vascular Flora of the Peninsula of Baja California, Mexico." *Journal of Vegetation Science* 18:327–36.

SEMARNAT-CONANP. 2010. Estrategia de cambio climático para áreas protegidas (ECCAP); SEMARNAT (Secretaria de Medio Ambiente y Recursos Naturales), CONANP (Comisión Nacional de Areas Naturales Protegidas): Mexico D.F., Mexico, 41.

Sheppard, P. R., A. C. Comrie, G. D. Packin, K. Angersbach, and M. K. Hughes. 2002. "The Climate of the Southwest." *Climate Research* 21:219–38.

Solomon, S., G. K. Plattner, R. Knutti, and P. Friedlingstein. 2009. "Irreversible Climate Change Due to Carbon Dioxide Emissions." *Proceedings of the National Academy of Sciences* 106:1704–9.

Spoerl, M. and J. C. Ravesloot. 1994. "From Casas Grandes to Casa Grande: Prehistoric Human Impacts in the Sky Islands of Southern Arizona and Northwestern Mexico." In *The Sky Islands of Southwestern United States and Northwestern Mexico*, L. F. Debano, G. J. Gottfried, R. H. Hamre, C. B. Edminster, P. F. Ffolliott, and A. Ortega-Rubio, eds. USDA Forest Service: General Technical Report RM-GTR-264, 6–18.

Suding, K. N. and D. Goldberg. 2001. "Do Disturbances Alter Competitive Hierarchies? Mechanisms of Change Following Gap Creation." *Ecology* 82:2133–49.

Thomas, C. D., A. Cameron, R. E. Green, M. Bakkenes, L. J. Beaumont, Y. C. Collingham, and B.F.N. Erasmus, et al. 2004. "Extinction Risk from Climate Change." *Nature* 427:145–48.

USFWS. 2011. *Landscape Conservation Cooperatives: Shared Science for a Sustainable Future.* www.fws.gov/science/shc/pdf/LCC_Fact_Sheet.pdf (accessed January 11, 2012).

Varady, R. and E. V. Ward. 2009. "Transboundary Conservation in Context: What Drives Environmental Change?" In *Conservation of Shared Environments: Learning from the United States and Mexico*, L. López-Hoffman, E. McGovern, R. G. Varady, and K. W. Flessa, eds. Tucson: University of Arizona Press, 9–23.

Warshall, P. 1994. "The Madrean Sky Island Archipelago: A Planetary Overview." In *The Sky Islands of Southwestern United States and Northwestern Mexico*, L. F. Debano, G. J. Gottfried, R. H. Hamre, C. B. Edminster, P. F. Ffolliott, and A. Ortega-Rubio, eds. USDA Forest Service: General Technical Report RM-GTR-264, 6–18.

Wenger, E. 1999. *Communities of Practice: Learning, Meaning and Identity.* Cambridge, UK: Cambridge University Press.

Wilder, M., C. A. Scott, M. Pineda Pablos, R. G. Varady, G. M. Garfin, and J. McEvoy. 2010. "Adapting across Boundaries: Climate Change, Social Learning, and Resilience in the US–Mexico Border Region." *Annals of the Association of American Geographers* 100:1–11.

Discussion Questions

1   Why is this area referred to as Sky Islands? What characterizes their topography that may be very different from where you live or the mountains that you visit?

2   What factors contribute to the high biodiversity of the Madrean Sky Islands?

3   What historical and present factors have contributed to threatening this biodiversity?

4   What is the primary biological concern of the authors from climate disruption in this region?

5   Find out what progress the Desert LCC and Sky Island Alliance have made in managing the Madrean Sky Islands relative to climate disruption and increasing border barriers.

CHAPTER 8

Yellowstone to Yukon, North America

Charles C. Chester, Jodi A. Hilty, and Wendy I. Francis

The Yellowstone to Yukon (Y2Y) region represents one of the best known and most advanced large landscape conservation efforts in the world. Due to the relatively high availability of data and to the region's numerous and diverse conservation groups and agencies, the regional response to climate change has been comparatively rapid—albeit much remains to be done. Ultimately, climate change preparedness will need to occur at local and subregional scales, with these efforts scaling up to support biodiversity conservation actions and policies across the Y2Y region. The Y2Y vision and on-the-ground efforts throughout the region constitute a working hypothesis that conservation at such a continental scale will enhance ecosystem resilience and provide opportunities for adaptation during this time of climate disruption.

Introduction to the Region

Mountain ecosystems around the world are significant for their rugged beauty as well as their biological and cultural significance. Over 3,200 kilometers (2,000 miles) long and half a million square miles in area, the Y2Y region represents one of the most intact mountain systems anywhere on the planet. All of the large carnivores and ungulates that were here in 1793—when Alexander Mackenzie became the first Caucasian to cross the North American continent north of Mexico—persist in this region. Approximately 10 percent of the Y2Y region lies under some form of protected-area status, including wildlife refuges, wilderness areas, and national, state,

Figure 8.1   A female grizzly and her cubs wander through the southern region of Y2Y. Grizzly bears are almost completely restricted to the Y2Y region in the lower forty-eight states and exist in dangerously low densities in some parts of southern Canada. Photo courtesy of WCS Jeff Burrell.

and provincial parks. Yet even as much of the region retains its ecological intactness, human activities have entailed substantial impacts, especially in the southern third of the Y2Y region. These activities threaten a number of high-profile species such as wolverines (*Gulo gulo*), wolves (*Canis lupus*), caribou (*Rangifer tarandus*), and grizzly bears (*Ursus arctos*) (Figure 8.1). The region's 700 protected areas, including the world's first national park (Yellowstone) and Canada's first national park (Banff), are key to maintaining the region's biodiversity (Figure 8.2).

Indigenous communities, including Native Americans in the United States and First Nations in Canada, have long occupied and traveled throughout these mountains. The Blackfoot people referred to the Rocky Mountains as "miistakis," or backbone of the world. As currently defined, the Y2Y region stretches across the traditional territory of thirty-one First Nations/Native American groups (Reeves 1998). Today tribes in the United States still operate as independent nations on reservations in various parts of the region and in some cases hold ceded rights (viz., access and even management authority over wildlife and lands beyond reservations). In Canada, southernmost First Nations often govern reserves and have treaty rights to pursue traditional activities on public lands. However, First Nations in British Columbia and in parts of Yukon and the Northwest Territories are still negotiating land settlements with federal and territorial governments.

With post-Columbian incursion and settlement of the Y2Y region, a new economy emerged that was largely based on various forms of natural resource extraction. These extractive industries—principally forestry, agriculture, ranching, grazing, mining, and oil and gas extraction—are extensive, meaning that they are pervasive on almost all but the most protected

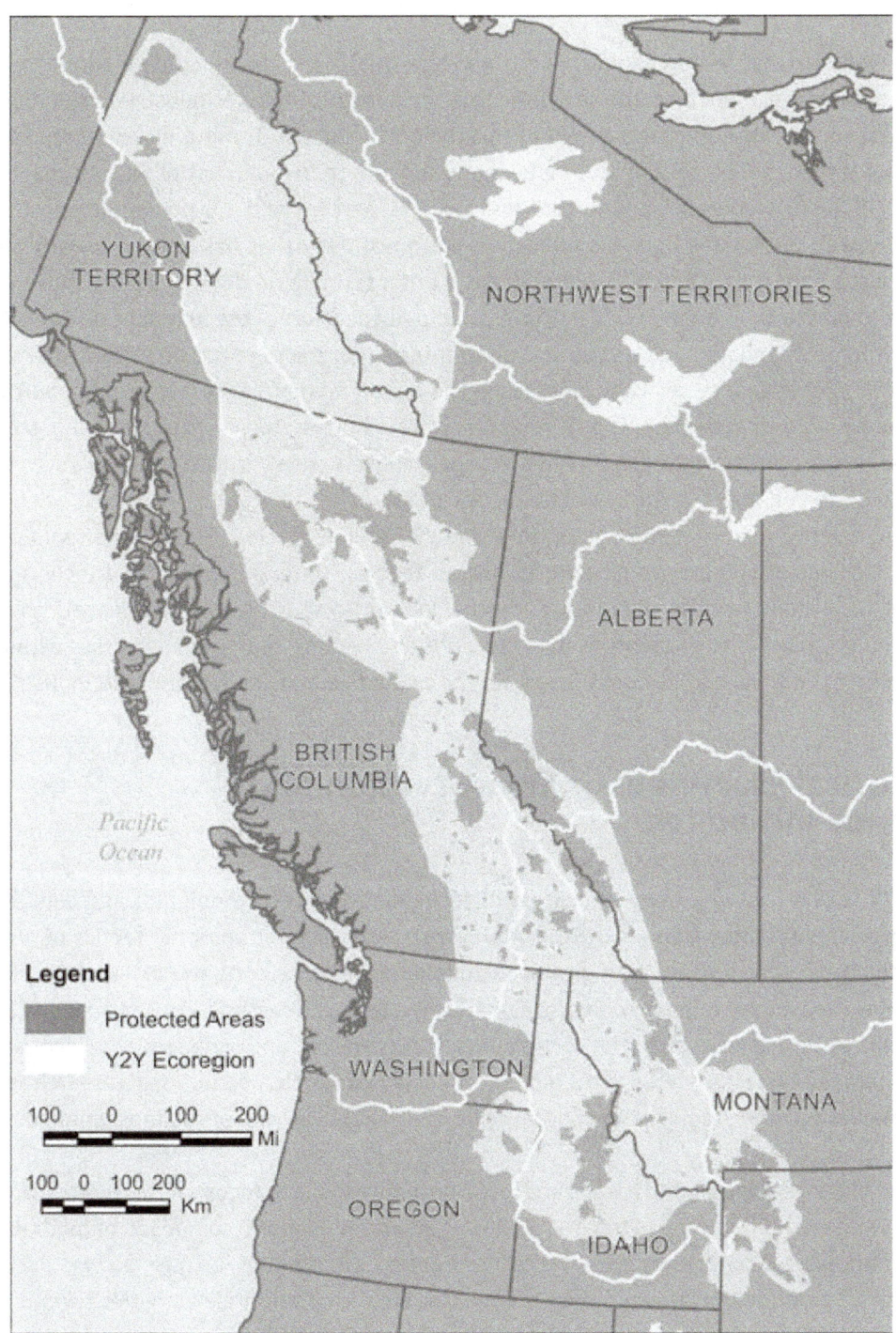

Figure 8.2   Major protected areas along the Yellowstone to Yukon landscape (Y2Y) in North America. Figure courtesy of WCS Andra Toivola.

lands. While natural resource extraction is an increasingly significant component of the economy as one travels north, in the last few decades nonconsumptive industries have begun to dominate much of the southern extent of this region. By the 1990s natural resources industries made up less than 6 percent of all employment and less than 5 percent of all personal income in the US Rocky Mountains, down from less than 11 percent and 9.6 percent, respectively, in 1969 (Rasker 1994). This figure is indicative of an economy that has shifted toward tourism, outdoor recreation, and "amenity migration," the latter term being defined as "the movement of people to places, permanently or part-time, principally because of the actual or perceived higher environmental quality and/or cultural differentiation of the destination" (Glorioso and Moss 2007, 138). Other service sector businesses in the region range from recreational opportunities such as skiing and snowmobiling to restaurants and hotels. Despite the changing economic base, however, politics in the region remain largely conservative with strong allegiances toward traditional stakeholders in the extractive industries.

Although the total human population in the Y2Y region is relatively low, the region's institutional complexity is labyrinthine in its myriad federal, state, provincial, and local agencies that manage land, wildlife, minerals, and other natural resources. In addition, self-governing indigenous groups in both countries manage lands and wildlife, regulate businesses within their jurisdictions, and manage businesses in both the extractive and service-oriented industries.

Historical Overview of Conservation and Science Initiatives

The Y2Y region has long been a focal point for hundreds of environmental and conservation nongovernmental organizations (NGOs). Along with NGOs using traditional tactics of advocacy and litigation, other conservation groups include land trusts, science and research institutions, sportsmen and angler organizations, local watershed associations, and local community groups. The Y2Y region has also been the focus of academic, agency, and independent scientists who have conducted an extensive array of biological, social, economic, and political research relevant to conservation in the region. As a consequence, Y2Y is arguably one of the most studied regions in the world.

The Y2Y Conservation Initiative is an NGO whose mission is to focus on the ecological health of the whole region. Formed in the mid-1990s as a "virtual network" composed of biologists and conservationists, a core component of the Y2Y Conservation Initiative's mission was to conceptualize and promote the larger landscape vision, as well as to prioritize activities toward critical areas of conservation concern across the region (Chester 2006). In addition to raising the profile of the Y2Y region as one of the first truly large landscape conservation efforts in the world, the organization has commissioned targeted science to identify and focus attention on a set of conservation priorities for places and issues, and has acted as a bridge helping partners share information and collaborate on shared visions or complementary efforts (for more background, see Y2Y 2011).

Since its inception, science has served as the core foundation of the Y2Y vision. Specifically, the rationale for the vision was derived from the scientific theories of island biogeography and metapopulation dynamics, both of which suggest that in the most general of terms, bigger protected areas closer together and embedded in a matrix of lands permeable to wildlife movement are more likely to increase the survival of individual species and thus to maintain the region's overall biodiversity. Research indicates that current protected areas in the Y2Y region are inadequate on their own to protect its biodiversity in the long term both because (1) they do not fully represent the region's biodiversity, and (2) they are not large enough to maintain healthy populations of wide-ranging species and the natural processes upon which they depend (Hilty et al. 2006). Such findings strongly support envisioning biodiversity conservation strategies at the scale of the Y2Y region, and of working toward both core area protection and connectivity between those core areas.

Threats to Biodiversity and the Response of Enhancing Connectivity

Threats to biodiversity in the Y2Y region run the familiar gamut, although the degree of threat shifts from south to north. Rural residential sprawl is the largest cause of land-use change in the southernmost stretches of the region, threatening many wildlife species (Hansen et al. 2002). As little as one house per section (a square mile) can turn the section from source (or good) habitat for grizzly bears to sink habitat where bears are likely to die as a result of conflict with humans (Schwartz et al., forthcoming). Farther north, areas that once experienced virtually no permanent human footprint are increasingly being explored and developed for various forms of natural resource extraction. Other threats range from roads and unsustainable forestry practices to intolerant attitudes toward some wildlife species. Some new categories of threat are looming, including the placement and extent of renewable energy projects and infrastructure, which represent a largely unknown threat. And of course, climate-related changes also threaten biodiversity.

The conservation community collectively uses a range of tools including private land conservation, wildlife restoration efforts, backcountry access management, various types of incentive programs, living with wildlife education and conflict reduction programs, and coordinated strategies that integrate core habitat protection campaigns. A major focus of conservation activities in the region is to target key public and private lands to maintain and enhance core areas and connectivity between core areas. Connectivity work in the Y2Y landscape has focused on mitigating the impacts of roadways such as the Trans-Canada Highway through Banff National Park and Highway 3 through the Crowsnest Pass area of southern Alberta (Ford et al. 2010; T. Lee et al. 2010) (Figure 8.3). Notably, the Y2Y region boasts the first federally designated migration corridor in the United States, the Path of the Pronghorn (Cohn 2010; ENS 2008). First documented by biologists from Grand Teton National Park and the Wildlife Conservation Society, the

discovery of this tenuous pathway inspired a collaborative effort by land management agencies
an

Figure 8.3 Across the Trans-Canada Highways bisecting Banff National Park, more than
twenty-five wildlife underpasses and overpasses have been built to enhance wildlife connectivity, including this overpass. Photo courtesy of WCS Jodi Hilty.

Collectively, the connectivity work at each of these scales supports the broader vision of the entire region functioning as an enormous corridor composed of protected core areas and linkages that ensure the long-term viability of all of the region's native species. While the Path of the Pronghorn represents an important step forward for corridor conservation, much work remains to be done in the broader Y2Y region.

Regional Effects of Climate Change

The Y2Y region currently hosts myriad research initiatives on the effects of climate change that range from studies on particular species in site-scale habitats to extensive large landscape analyses. A few examples of broad-scale research initiatives include the following:

- The Pacific Climate Impacts Consortium, which focuses on climate change impacts in the Pacific Northwest, including the portion of British Columbia that lies within the Y2Y region (http://pacificclimate.org);

- The Climate Adaptation Secretariat of the Province of British Columbia, which provides advice to government and reports on actions to implement a climate action plan (http://www.env.gov.bc.ca/cas);

- The US Forest Service Rocky Mountain Research Station, which is examining the effects of climate change on water supply and quality, wildland fire, and terrestrial ecosystems (RMRS 2009);

- The Northwest and North Central Climate Science Centers, established by the US Geological Survey, which "synthesize existing climate-change-impact data and management strategies, help resource managers put them into action on the ground, and engage the public through education initiatives" (NCCSC 2010; USDOI 2010); and

- The Department of the Interior's Great Northern Landscape Conservation Cooperative, a goal of which is to foster conservation science-management partnerships to provide scientific and technical support for conservation at landscape scales under conditions of climate change and other landscape stressors (http://www.nrmsc.usgs.gov/gnlcc).

Such efforts demonstrate that both academic scientists and government officials have begun to take steps to understand the impacts of climate change across the Rocky Mountains and adjacent ecoregions. These and other science efforts were summarized in *Moving toward Climate Change Adaptation: The Promise of the Yellowstone to Yukon Conservation Initiative for Addressing the Region's Vulnerabilities* (Graumlich and Francis 2010). Much of the following discussion summarizes this report's assessment and review of climate impacts in the Y2Y region.

Scientists have long recognized the potential for rapid climate change in the Rocky Mountains (Luckman 1990). In an extensive study on climate change in the US portion of the

Rocky Mountains, Reiners et al. (2003, 178) noted that "[s]hifts in environmental gradients across Rocky Mountain landscapes are likely to lead to rapid growth in the numbers of recognizably threatened species." Today, a rapidly expanding body of scientific evidence indicates that the Y2Y region is experiencing climate change impacts and that such continued changes will significantly alter the native biota. Mean annual temperatures have been increasing throughout the region, and scientists have documented changing precipitation patterns and an increased frequency of extremes in comparison with historical records. Although average temperature changes are relatively small, it is the increase in extreme temperatures that could have the greatest impact. This is because many species exist within relatively narrow climate envelopes, and extreme temperatures can lead to stress, reproductive failure, and even death. As Graumlich and Francis (2010, 28) put it, there is strong evidence that:

> ... the climates of the Y2Y region have already changed beyond the limits of historic variation; that these climatic changes are having ecological impacts; that continued changes, especially warming, will have long-term, unprecedented future impacts; and that landscape-scale conservation is a central element of limiting and adapting to such inevitable changes.

Winters are shorter and warmer in the twenty-first century than in previous decades, and precipitation is increasingly falling as rain rather than snow, the effects of which include changes in seasonal water availability and glacier endurance. Glaciers are literally disappearing, and Glacier National Park could soon be a misnomer, a fact that has made the park a prominent "poster-child" of climate change (Hall and Fagre 2009). Those species that depend on the presence of snow and ice on a year-round basis will experience the greatest impacts. For example, because wolverines typically depend on cool places both to cache their food where it will not rot, and to den in areas where deep snow provides insulation, the loss of permanent snow could be problematic for this species (Chadwick 2010; Copeland et al. 2010; McKelvey et al. 2011; Peacock 2011). In addition to changes in climatic and physical variables, scientists are also documenting phenological alterations (such as changes in hibernation patterns) and shifts in species distributions. Ecosystem function can also depend on phenological timing between codependent species, which can decouple as such species seek to adapt to changing conditions (Graumlich and Francis 2010).

Some species may benefit from climate change. Yellow-bellied marmots (*Marmota flaviventris*) in the Rockies are gaining prehibernation weights much earlier in the summer and then exiting hibernation earlier in the spring, leading to their population expanding beyond historical numbers. However, researchers have posited that such benefits may be short lived due to the long-term probability of extended drought (Ozgul et al. 2010). More broadly, both observational data and modeling indicate that species' distributions are changing and are likely to further change over the next century, and that those species that cannot move to areas with desirable climate and habitat conditions are likely to perish (Graumlich and Francis 2010).

Arguably the most visible impacts of climate change in the Y2Y region are the widespread outbreaks of pine beetles (mostly the mountain pine beetle, *Dendroctonus ponderosae*). Although large infestations have occurred in the past, researchers have characterized the extent and longevity of recent infestations as unprecedented, particularly inasmuch as warmer conditions at higher elevations have brought the beetles into contact with five-needle pine species, such as whitebark pine (*Pinus albicaulis*) (Carroll et al. 2006; Logan et al. 2010; Robertson et al. 2009; Taylor and Carroll 2004). Cold winters have historically limited pine beetle infestations, and the extensive infestations seen on the landscape today result from milder winters and, at high altitudes, the pine beetle's corresponding ability to shorten its life cycle and thus increase its reproductive rate (Bentz et al. 2010). Extensive and rapidly expanding stands of dead and dying trees are increasingly ubiquitous throughout much of the Y2Y region.

Approaches to Conservation under Climate Change

Conservationists in the Y2Y region have long recognized the threat of climate change. For example, when the Y2Y Conservation Initiative was founded in 1993, one of the rationales for envisioning conservation across such a large scale was the need to conserve landscapes at sufficient scale to be robust to climate change (Harvey 1998). Yet as evinced by the large-scale regional climate prediction and adaptation efforts covering large portions of the Y2Y region discussed earlier, climate change is emerging as a new focus for many conservationists in the region.

Not surprisingly, government agencies, communities, conservation NGOs, and other entities in the region vary widely in the terms of their engagement on climate change. Some have continued their current line of work and have not added any emphasis or focus to climate change. Many local land trusts, for example, have a fairly narrow remit, and with the exception of a few instances where water conservation is beginning to take a higher profile, their business strategy and programmatic approach is often relatively unaffected by climate change. Other entities indicate that their conservation priorities and actions already *are* a response to climate change, inasmuch as their focus on protecting core areas, maintaining connectivity, and removing or minimizing nonclimate stressors constitute their priorities for ensuring that ecosystems are resilient to and/or are able to adapt to climate change (Graumlich and Francis 2010; Hansen et al. 2010). For example, groups collaborating to restore and maintain the last tendrils of connectivity in the transboundary Cabinet-Purcell region describe their work as both a strategy to conserve the region's grizzly bears and a tool for enhancing the region's resilience to climate change (Y2Y 2011).

Other entities have reevaluated their biodiversity conservation or management goals in light of climate change. For the Crown of the Continent Ecosystem, the Wildlife Conservation Society (WCS) analyzed roadless areas for their value to wildlife, a component of which focused on connectivity and climate change to guide protected area design (Weaver 2011). This work addresses the difficult question: if we cannot protect all of the current roadless areas, which ones might be most important to conserve, given climate change and connectivity needs? Likewise, targeted beaver (*Castor canadensis*) restoration efforts in Alberta by a local watershed group are part of a larger effort to raise the profile of water under climate disruption and of beaver as a component of free ecosystem services that have the capacity to help retain mountain water runoff for longer periods, cool in-stream water, and raise ground water levels (WCS Canada 2009).

The reality is that many institutions throughout the Y2Y region are still working to incorporate climate change considerations at the planning level. Management plans for the US Forest Service, for example, are supposed to consider the impacts of climate change, but how this translates to on-the-ground natural resource decisions is less clear. Similarly, western states are working to incorporate connectivity and climate change into their state wildlife action plans to help them map out their priorities into the future (WGA 2008; AFWA 2009). Other efforts focus on bringing climate science information to local stakeholders, helping them make robust decisions under uncertain climatic conditions. For example, for both the Greater Yellowstone Ecosystem and the transboundary US–Canada region, the US Fish and Wildlife Service and WCS collaborated on two workshops to discuss climate change impacts on grizzly bears and wolverines, to identify strategies to conserve these species given climate change, and to enhance communication and collaborative action among regional stakeholders (Cross and Servheen 2009; Servheen and Cross 2010). Farther north, the Yukon Conservation Society has an Energy and Climate Change Program that has developed a proposal for a "Yukon Carbon Fund" (Taggart 2009). On-the-ground examples of management decisions that consider the impacts of climate change are limited, but if trends continue, this will likely become a common land management practice throughout the Y2Y region.

Roadblocks and Opportunities

Perhaps the most difficult challenge facing conservationists in the Y2Y region is their own history—specifically, their longstanding tradition of protecting biodiversity by staking out specific areas for protection of *in situ* resources, a pattern originating with the 1872 establishment of Yellowstone National Park. Although Yellowstone was originally intended to protect its rare geological features, it soon became as much a sanctuary for wildlife that was rapidly disappearing from a developing continent. The working assumption was that if humankind were to place a sufficient portion of land under conservation status, protected areas could play a key role in conserving a region's biodiversity. Although this is still the case, that role is shifting; climate change now demands that conservationists work across much larger landscapes, both to protect the ability of species to move across those landscapes and to preserve ecosystem functions rather than just their constituent components. Precedent, not to mention mechanisms, are lacking for effective and coordinated multijurisdictional planning and implementation.

Another major challenge is that many citizens reject the premise of anthropogenic climate change. Despite overwhelming scientific evidence, this backlash influences particular sectors in the Y2Y region and presents significant challenges in incorporating climate change considerations into policy and management decisions. Also, the region lacks new resources to retain or expand climate expertise either on the science or policy side. This is particularly true in government agencies, where a mandate for climate change adaptation can translate into additional work without new resources. Further, efforts to reduce greenhouse gas emissions mean an increased demand for renewable energy sources, and growing wind and hydrogenerated electricity are already showing an increasing footprint in the Y2Y region. As these expand, they may come into conflict with conservation priorities and could decrease the region's intactness and even its resilience to climate change. Examples are the proposed Site C hydroelectric dam on the Peace River in northern BC and the hundreds of independent power projects under consideration for streams and rivers throughout the BC portion of the Y2Y region (Evenden 2009).

Finally, stakeholders have not yet fully conceptualized the problem of rapid climate change or its solutions. There is a lack of agreement around whether the goal is to resist the impacts of climate change or to enable transformative ecological changes that sustain changing configurations of biodiversity. For example, should a waterfowl reserve with drying wetlands begin to pump water to maintain the same conditions? If so, under what circumstances? When will pumping become too costly or ineffective? When should the focus instead be on managing for change?

Conclusion and Recommendations

Given the significant and unpredictable ecosystem changes that will inevitably accompany alterations in temperature and precipitation, a business-as-usual conservation model will not protect the biodiversity of the Y2Y region. For example, the availability of water is already a contentious issue in much of the region. As water reserves and the timing of peak season flows change, securing adequate supplies of water for all users is likely to become one of the most contentious conservation issues in the twenty-first century. Changes in precipitation quantity and type (viz., rain or snow) in the Y2Y region mean that certain seasons may be more arid, resulting in higher agricultural and municipal demand during certain times of year. Pressure to construct dams as a means to prepare for times of drought is likely to build. At the same time, species dependent on free-flowing and cool streams will almost certainly be at odds with such demands. Conservationists in the Y2Y region need to prioritize this issue and design proactive solutions before it becomes too late. Likewise, ensuring connectivity for those species that have the capacity to move will be increasingly important, and consequently it will be essential to understand what kinds of corridors or linkages may be robust over time (and for which species) given climate change (Cross et al. in press).

In addition, it may be useful to identify the likely locations of refugia from climate change and to prioritize these areas for protection (Graumlich and Francis 2010). During the last ice age, the Y2Y region contained several refugias where glaciation did not occur. Although refugias in a warming future will likely be quite different than those of a glaciated past, the varied terrain of the Y2Y landscape harbors conditions that could provide species with refuge from the current and looming changes.

The Y2Y region is fortunate inasmuch as it has attracted many scientists over the past century, and thus we have a significant amount of historic and current science to guide biodiversity conservation efforts under conditions of climate change. Yet even with this knowledge, considerable uncertainty remains in terms of predicting a climate change trajectory within any subregion of the landscape. Despite such uncertainties, conservation activities must ensure that the Y2Y region can be a place where plants and animals can shift to find appropriate niche space as habitats change.

The Y2Y region arguably represents one of the planet's best opportunities to work at a large landscape scale. As one of the world's most intact mountain ecosystems, the Y2Y region's ecological integrity makes it more resilient to climate change than many other regions. As conditions change, dispersal-limited species are likely to find more of an opportunity to move to appropriate climatic spaces due to the diversity of climate niches within the vast Y2Y region—niches that result from the region's relative intactness, its wide latitudinal expanse, and its mountainous terrain that incorporates both aspect (north-east-southwest) and elevational gradients (Chen et al. 2011). Given the inherent resilience and adaptability of the Y2Y region, it is all the more important to implement sound conservation actions now to ensure that this region can be robust for species conservation over the coming centuries.

References

AFWA (Association of Fish and Wildlife Agencies). 2009. *Voluntary Guidance for States to Incorporate Climate Change into State Wildlife Action Plans and Other Management Plans.* Washington, DC: Climate Change Wildlife Action Plan Work Group, AFWA.

Bentz, B. J., J. Régnière, C. J. Fettig, E. M. Hansen, J. L. Hayes, J. A. Hicke, R. G. Kelsey, J. F. Negrón, and S. J. Seybold. 2010. "Climate Change and Bark Beetles of the Western United States and Canada: Direct and Indirect Effects." *BioScience* 60:602–13.

Carroll, A. L., J. Régnière, J. A. Logan, S. W. Taylor, B. J. Bentz, and J. A. Powell. 2006. *Impacts of Climate Change on Range Expansion by the Mountain Pine Beetle.* Victoria, British Columbia: Pacific Forestry Centre, Canadian Forest Service, Natural Resources Canada. Mountain Pine Beetle Initiative Working Paper, 2006–14.

Chadwick, D. H. 2010. *The Wolverine Way.* Ventura, CA: Patagonia.

Chen, C., J. K. Hill, R. Ohlemüller, D. B. Roy, and C. D. Thomas. 2011. "Rapid Range Shifts of Species Associated with High Levels of Climate Warming." *Science* 333:1024–26.

Chester, C. C. 2006. *Conservation across Borders: Biodiversity in an Interdependent World.* Washington, DC: Island Press.

Cohn, J. P. 2010. "A Narrow Path for Pronghorns." *BioScience* 60:480.

Copeland, J. P., K. S. McKelvey, K. B. Aubry, A. Landa, J. Persson, R. M. Inman, and J. Krebs. 2010. "The Bioclimatic Envelope of the Wolverine (*Gulo gulo*): Do Climatic Constraints Limit Its Geographic Distribution?" *Canadian Journal of Zoology* 88:233–46.

Cross, M. S., J. A. Hilty, G. M. Tabor, J. J. Lawler, L. J. Graumlich, and J. Berger. In press. "From Connect-the-dots to Dynamic Networks: Maintaining and Enhancing Connectivity as a Strategy to Address Climate Change Impacts on Wildlife." In *Wildlife Conservation in a Changing Climate*, J. Brodie, D. Doak, and E. Post, eds. Chicago: University of Chicago Press.

Cross, M. S. and C. Servheen. 2009. *Climate Change Impacts on Wolverines and Grizzly Bears in the Northern U.S. Rockies: Strategies for Conservation*. Final Workshop Summary Report October 6–7, 2009. Wildlife Conservation Society and U.S. Fish and Wildlife Service. May 24, 2010.

ENS (Environmental News Service). 2008. *Ancient Pronghorn Path Becomes First U.S. Wildlife Migration Corridor*. ENS. June 25. www.ens-newswire.com/ens/jun2008/2008-06-17-091.html (accessed January 11, 2012).

Evenden, M. 2009. "Site C Forum: Considering the Prospect of Another Dam on the Peace River." *BC Studies* 161:93–114.

Ford, A. T., A. P. Clevenger, and K. Rettie. 2010. "The Banff Wildlife Crossings Project: An International Public-Private Partnership." In *Safe Passages: Highways, Wildlife, and Habitat Connectivity*, J. P. Beckmann, A. P. Clevenger, M. P. Huijser, and J. A. Hilty, eds. Washington, DC: Island Press, 157–72.

Glorioso, R. S. and L. A. G. Moss. 2007. "Amenity Migration to Mountain Regions: Current Knowledge and a Strategic Construct for Sustainable Management." *Social Change* 37:137–61.

Graumlich, L. and W. L. Francis. 2010. *Moving toward Climate Change Adaptation: The Promise of the Yellowstone to Yukon Conservation Initiative for Addressing the Region's Vulnerabilities*. Canmore, Alberta: Yellowstone to Yukon Conservation Initiative. www.y2y.net/data/1/rec_docs/898_Y2Y_Climate_Adaptation_Report__FINAL_Web.pdf (accessed January 6, 2012).

Hall, M. H. P. and D. B. Fagre. 2009. "Modeled Climate-induced Glacier Change in Glacier National Park, 1850–2100." *BioScience* 53:131–40.

Hansen, A. J., R. Rasker, B. Maxwell, J. J. Rotella, J. D. Johnson, A. Wright Parmenter, U. Langner, W. B. Cohen, R. L. Lawrence, and M. P. V. Kraska. 2002. "Ecological Causes and Consequences of Demographic Change in the New West." *Bioscience* 52:151–62.

Hansen, L., J. Hoffman, C. Drew, and E. Mielbrecht. 2010. "Designing Climate-smart Conservation: Guidance and Case Studies." *Conservation Biology* 24:63–69.

Harvey, A. 1998. *A Sense of Place: Issues, Attitudes and Resources in the Yellowstone to Yukon Ecoregion*. Canmore, AB, Canada: Yellowstone to Yukon Conservation Initiative.

Hilty, J. A., W. Z. Lidicker, and A. M. Merenlender. 2006. *Corridor Ecology: The Science and Practice of Linking Landscapes for Biodiversity Conservation*. Washington, DC: Island Press.

Lee, T., M. Quinn, and D. Duke. 2010. "A Local Community Monitors Wildlife along a Major Transportation Corridor." In *Safe Passages: Highways, Wildlife, and Habitat Connectivity*, P. B. Jon, A. P. Clevenger, M. P. Huijser and J. A. Hilty, eds. Washington, DC: Island Press, 277–292.

Logan, J. A., W. W. Macfarlane, and L. Willcox. 2010. "Whitebark Pine Vulnerability to Climate-driven Mountain Pine Beetle Disturbance in the Greater Yellowstone Ecosystem." *Ecological Applications* 20:895–902.

Luckman, B. H. 1990. "Mountain Areas and Global Change: A View from the Canadian Rockies." *Mountain Research and Development* 10:183–95.

McKelvey, K., J. Copeland, M. Schwartz, J. Littell, K. Aubry, J. Squires, S. Parks, M. Elsner, and G. Mauger. 2011. "Climate Change Predicted to Shift Wolverine Distributions, Connectivity, and Dispersal Corridors." *Ecological Applications* 21:2882–97.

NCCSC (North Central Climate Science Center). 2010. *North Central Climate Science Center*. U. S. Geological Survey. www.doi.gov/csc/northcentral/about.cfm (accessed January 10, 2012).

Ozgul, A., D. Z. Childs, M. K. Oli, K. B. Armitage, D. T. Blumstein, L. E. Olson, S. Tuljapurkar, and T. Coulson. 2010. "Coupled Dynamics of Body Mass and Population Growth in Response to Environmental Change." *Nature* 466:482–85.

Peacock, S. 2011. "Projected 21st Century Climate Change for Wolverine Habitats within the Contiguous United States." *Environmental Research Letters* 6:014007.

Rasker, R. 1994. "A New Look at Old Vistas: The Economic Role of Environmental Quality of Western Public Lands." *University of Colorado Law Review* 65:369–99.

Reeves, Brian O. K. 1998. "Sacred Geography: First Nations of the Yellowstone to Yukon." In *A Sense of Place: Issues, Attitudes and Resources in the Yellowstone to Yukon Ecoregion*, Ann Harvey, ed. Canmore, Alberta: Yellowstone to Yukon Conservation Initiative, 31–50.

Reiners, W. A., J. S. Baron, D. M. Debinski, S. A. Elias, D. B. Fagre, J. S. Findley, and L.O. Mearns. 2003. "Natural Ecosystems 1: The Rocky Mountains." In *Rocky Mountain/Great Basin Regional Climate-change Assessment*, F. H. Wagner, ed. Logan: Utah State University report for the U. S. Global Change Research Program, 145–84.

RMRS (Rocky Mountain Research Station). 2009. *2009 Climate Change Research Strategy*. U.S. Forest Service, RMRS. www.fs.fed.us/rmrs/docs/climate-change/climate-change-research-strategy.pdf (accessed January 10, 2012).

Robertson, C., T. A. Nelson, D. E. Jelinski, M. A. Wulder, and B. Boots. 2009. "Spatial–temporal Analysis of Species Range Expansion: The Case of the Mountain Pine Beetle, *Dendroctonus ponderosae*." *Journal of Biogeography* 36:1446–58.

Schwartz, C. C., P. H. Gude, L. Landenburger, M. A. Haroldson, and S. Podruzny. Forthcoming. "Impact of Rural Residential Development on Grizzly Bear Habitat in the Greater Yellowstone Ecosystems." *Journal of Wildlife Management*.

Servheen, C. and M. S. Cross. 2010. *Climate Change Impacts on Grizzly Bears and Wolverines in Northern US and Transboundary Rockies: Strategies for Conservation*. Report on a workshop held September 13–15, 2010, in Fernie, British Columbia.

Taggart, M. 2009. *The Feasibility of a Yukon Carbon Offset Fund*. Marsh Lake, Yukon: Research Northwest for the Yukon Conservation Society. March 27. www.yukonconservation.org/energydocuments/The_Feasibility_of_a_Yukon_Carbon_Fund_FINAL_REPORT_2009_03_27-1%5B1%5D.pdf (accessed January 11, 2012).

Taylor, S. W. and A. L. Carroll. 2004. "Disturbance, Forest Age, and Mountain Pine Beetle Outbreak Dynamics in BC: A Historical Perspective." In *Mountain Pine Beetle Symposium: Challenges and Solutions, October 30–31, 2003, Kelowna, British Columbia, Canada*, T. L. Shore, J. E. Brooks, and J. E. Stone, eds. Victoria, BC: Pacific Forestry Centre, 41–51.

USDOI (United States Department of Interior). 2010. *Climate Change*. U.S. Department of the Interior. www.doi.gov/whatwedo/climate/index.cfm (accessed January 11, 2012).

WCS Canada (Wildlife Conservation Society Canada). 2009. *On the Golden Pond*. Pamphlet. www.wcscanada.org/LinkClick.aspx?fileticket=VUpxOxePLdo%3dandtabid=3071 (accessed January 11, 2012).

Weaver, J. 2011. "Conservation Value of Roadless Areas for Vulnerable Fish and Wildlife Species in the Crown of the Continent Ecosystem, Montana." Working Paper, no. 40. April.

WGA (Western Governors' Association). 2008. *Western Wildlife Habitat Council Established*. Jackson, Wyoming: WGA. June 29. www.westgov.org/wga/publicat/wildlife08.pdf (accessed January 11, 2012).

Y2Y (Yellowstone to Yukon). 2011. *Cabinet-Purcells Mountain Corridor: Collaborative Projects*. Yellowstone to Yukon Conservation Initiative. www.y2y.net/Default.aspx?cid=4-67-532andpre=view (accessed January 11, 2012).

Discussion Questions

1   Why is the Yellowstone to Yukon (Y2Y) region so appealing for species conservation?

2   What historical and present land uses conflict with species conservation in this region?

3   Describe how climate disruption is already affecting two species.

4   What strategies are needed for species conservation in Y2Y?

5   What are the described challenges in making progress? Try to determine if progress has been made in the past several years.

CHAPTER 9

Climate Change and Biodiversity in the Great Lakes Region

From "Fingerprints" of Change to Helping Safeguard Species

Kimberly R. Hall and Terry L. Root

O ver the last century, the average global surface temperature has increased approximately 0.8°C, and the rate of warming continues to accelerate (Trenberth et al. 2007). Even with this amount of warming, which is small compared to the net increase we may see in the relatively near future (an additional 1.1° to 6.4°C or more increase in the global average by 2100 according to Meehl et al. 2007), wild species are already exhibiting discernible changes (Root et al. 2003; Parmesan and Yohe 2003; Parmesan 2006). Like other regions at moderate latitudes, temperature change projections for the Great Lakes region are somewhat higher than projections for the global average. Temperatures in winter are rising faster than any other season, which may seem like a welcome change to some residents, but can lead to costly impacts in agricultural and managed forest systems through increases in the survival of crop and forest pests (CCSP 2009; Bradshaw et al., this volume). Specifically, winter temperatures will be "less cold" (nightly low temperatures are expected to increase more than the daytime highs), contributing to lengthening of the frost-free growing season, which has already increased by more than one week (Field et al. 2007; CCSP 2009; Andresen, this volume). By the end of this century, average summer temperatures are projected to increase by 3° to 7°C, leading to dramatic increases in the frequency of heat waves, especially if the temperature increase is at the higher end of this range (Christensen et al. 2007; CCSP 2009; Hayhoe et al. 2010b). Summer surface water temperatures in the upper Great Lakes (Michigan, Huron, Superior) are currently increasing even faster than the air temperatures, and these changes are triggering a whole range of system-wide impacts, including increases in wind and current speeds, and increases in the duration of the stratified period (Austin and Coleman 2007, 2008; Desai et al. 2009; Dobiesz and Lester 2009).

The rate at which these temperature changes are occurring suggests that many, if not most, wild species will experience climate change as a stressor that reduces survival and/or reproduction, and thus has strong potential to lead to population declines, or even extinction. The most recent Intergovernmental Panel on Climate Change (IPCC) report, which represents the consensus view of a team of hundreds of scientists from across the globe, suggests that 15 to 40 percent of known species will be at an increasingly high risk of extinction as global mean temperatures reach 2° to 3°C above preindustrial (or 1.2 to 2.2°C above current) levels (Field et al. 2007, based on work in Thomas et al. 2004). Identifying the most vulnerable species, and what we can do to safeguard their populations, is a major challenge, and each day that passes without reductions in the gases that trap heat in our atmosphere (carbon dioxide, methane) stacks the odds further against the survival of sensitive species and systems.

Here we review observed and predicted impacts of climate change on biodiversity in the Great Lakes region, with the goal of helping to inform and motivate actions that reduce emissions, and to promote actions that help safeguard species and systems so that they can adapt to ongoing changes. Although the scientific literature documenting climate change impacts on species is rapidly expanding, predicting how and when focal species will respond is a daunting task, in part because changes in temperature and other factors are happening within a climate system that already exhibits high natural variation across both space and time. The observed changes, along with ecological theory, allow us to develop "rules of thumb" for how species are likely to respond to the most direct aspects of climate change (e.g., changes in air or water temperature). In addition, experimental studies or predictive models may provide clues as to how several climate factors (temperature, precipitation pattern) may interact. However, it is important to recognize from the start that because many climate factors and interacting species are changing simultaneously, species may show very complex responses, and thus it is often very hard to categorize relative risk. Predictions of responses to change and the relative vulnerability of species become even more uncertain when we try to put them in the context of all of the other stressors that wild species and ecosystems currently face—such as habitat loss, invasion by non-native species, changes in hydrology, and pollution (see Mackey, this volume)—and changes they will face in the future, including actions that societies take in response to changes in climate.

Understanding how climate change will impact species is further complicated by the fact that several aspects of climate change involve feedback loops, or have the potential to lead to tipping points (dramatic shifts in a system in response to an incremental change in some climate factor). For example, surface water temperatures of the upper Great Lakes are showing summer temperature increases that exceed regional temperature increases on land, in part due to positive feedbacks on the warming rate due to reductions in ice cover. Specifically, ice reflects energy from the sun and insulates the water from the warming air, but melts more quickly when the air is warmer, and this loss of ice cover accelerates the rate of surface water warming (Austin and Colman 2007, 2008; Dobiesz and Lester 2009). Study of lakes also provides a good example of a critical temperature threshold, as areas that are deep enough undergo summer stratification, which is triggered by warming of the surface layer above about 4°C (McCormick and Fanenstiel 1999). When bodies of water stratify, they separate into a warmer, oxygen-rich top layer (the epilimnion), a zone of rapid temperature change (the thermocline), and a colder bottom layer (the hypolimnion), where decomposition of dead plant and animal matter is a dominant process, and oxygen can become limiting. To make a complex story even more so, recent research demonstrates that the fact that the lakes are warming faster than the air is also important to the process of stratification in large lakes, as decreases in the air-to-lake temperature gradient leads to increases in surface wind speeds, which increases currents and acts in combination with the temperature to further lengthen the stratified period (Desai et al. 2009).

Why are we concerned about how climate change factors may drive changes in the timing or duration of stratification? To put it simply, changes in temperature, both direct and through the ice and wind-related mechanisms described above, have the potential to profoundly change how large lakes in our region function. The differences in temperature, light availability, and other factors that occur as a result of stratification provide a diversity of habitats within stratified lakes, which allows species with a wide variety of temperature and other habitat requirements to persist. The timing of stratification, as well as the timing of the fall "turnover," when the oxygen-rich surface waters cool and increase in density, and finally sink down and mix with the others, can be a critical factor influencing the viability of lake species, especially cold-water fish. Given that changes in temperatures for the upper Great Lakes are projected to continue to match or exceed the air-temperature increases, we should expect to see longer stratified periods and increased risk of oxygen deficits below the thermocline in late summer (Magnuson et al. 1997; Jones et al. 2006; Dobiesz and Lester 2009). Increases in the duration of the stratified period of over two weeks have already been observed for Lake Superior (Austin and Colman 2008), and projections for the end of this century suggest that we could see lakes stratify for an additional one and a half months (Lake Erie for a lower emissions scenario and thus less climate change) to three months (Lake Superior under the assumption of higher future emissions; Trumpickas et al. 2009). As the depth and latitude of a lake, lake basin, or bay decreases, it is less likely to show stratification, but some shallow-water bodies will exhibit oxygen-poor "dead zones" because shallow water warms more rapidly, and warmer water holds less oxygen and leads to increases in respiration rates for aquatic species. As warming continues, we should expect more and more areas to develop "dead zones," and for others to transition from stratifying in summer to not stratifying at all, with resultant loss of species that depend on habitats characterized by colder water (see Mackey, this volume, for more discussion of the causes and impacts of dead zones).

Many of us recognize that the Great Lakes moderate temperature changes in nearshore areas within our region, and as a result, these huge water bodies may be perceived as being somewhat protected from the impacts of changes in climate. The fact that at least the upper lakes are warming and changing faster than most land areas in our region may be surprising to many, and this fact challenges us to change how we think about these dynamic, critically important ecosystems. We argue that this well-documented "surprise" of rapidly warming large lakes is likely to be one of many, and we must be prepared for surprises that will change ecological conditions such that the knowledge we have relied upon to inform management and conservation practices may no longer be relevant. First and foremost, we need to act to minimize the degree of change, which means we need to effectively and efficiently reduce our emissions. Yet, as is clear from the examples in this and many other chapters in this volume, we are already seeing the impacts of climate change in the Great Lakes region, and so we also need to help people adapt, and safeguard the species and ecosystems upon which we depend. To move forward on adaptation, we will need to anticipate changes, learn from the results of our actions, and be prepared to quickly change course in response to the unexpected. The first steps in taking action to protect species are to review what we already know, and to see how ecological theory can help us understand what changes in species and ecosystems are likely.

Our review of the impacts of climate change on wild species begins with and focuses on responses to increases in air and water temperature. We spend the most time on temperature, rather than other key factors like precipitation patterns, lake-level changes, or direct effects of increased CO_2 concentrations, for three reasons. First, mean annual temperature has shown a strong and consistent pattern of increase, and projections from various Global Circulation Models (GCMs) show strong agreement in their predictions of increasingly rapid temperature rise (Meehl et al. 2007). Second, changes in many species can be statistically attributed directly to the anthropogenic (human-caused) component of this temperature increase (Root et al. 2005; Rosenzweig et al. 2007), and further temperature changes will take place in the context of geographic patterns that are well understood (i.e., air temperatures cool as you go toward the polar regions or up in altitude; in summer, deep water is cooler than surface water). Third, there is a well-developed body of ecological theory that helps us to frame predictions of how species will respond to increases in the various facets of temperature, such as warmer summers, longer growing seasons, and reductions in the number of days with temperatures below freezing. Often, there are many mechanisms through which the same general impact is realized: for example, warmer waters can stress species because the increase in temperature reduces the oxygen-holding capacity of water, and because at higher temperatures, the respiration rate of many species, which determines how much oxygen is needed, is higher.

Quantifying responses of wild species to climate change is a major challenge, as shorter-term (e.g., monthly, yearly) variability in temperature, as well as the wide variety of other factors influencing species, leads to lots of "noise" in our scientific data. Thus, to detect directional patterns within data sets of species and temperature changes, we rely on long-term data sets (e.g., over 20 years) to evaluate and document wild species' responses. Just a decade or two ago, such data sets were relatively rare in the published literature, but as awareness of climate change has grown, data originally collected for a wide variety of different purposes have been linked with temperature for evaluating the responses of species and ecological systems to climate changes (Parmesan 2006). These long-term studies, a small subset of which are described below, help us to understand current stresses on species and current differences across species groups in terms of type and degree of response, and help us to anticipate future responses as the climate continues to change.

Overview of Observed Changes

Responses already observed in wild animals and plants to warming temperatures can be grouped into five basic types: (1) spatial shifts in ranges boundaries (e.g., moving north in the Northern Hemisphere); (2) spatial shifts in the density of individual animals or plants within various sub-sections of a species' range; (3) changes in phenology (the timing of events), such as when leaves emerge in spring, or when birds lay their eggs; (4) mismatches in the phenology of interacting species; and (5) changes in genetics. These categories are not mutually exclusive—for example, a change in the timing of bird migration can represent both a phenological shift and a shift in gene frequencies (genetics). Further, it is important to recognize that shifts in density and abundance include extirpation (loss of a species from a local area) and extinction. The categorization depends primarily on the organizational level (populations, individuals, genes) at which the relationship between a species and its environment were studied. As temperatures continue to rise, and our understanding of other types of climate change impacts improve, we can expect that the volume and variety of impacts on wild species attributable to climate change will continue to increase rapidly as well.

A majority of wild species show predictable changes in responses to increasing temperatures, and the role of temperature in shaping species' life histories is strong. In other words, temperature regime is a key element to which species have adapted over long (evolutionary) time periods. The strong role that temperature plays can be observed in similarities between the edges of species ranges and mapped temperature patterns, north-south patterns in body size within the same species (northern animals are typically a bit larger), and geographic variation in seasonal activities like hibernation and timing of breeding (Root 1988; Brown 1995; Millien et al. 2006). The potential effects of temperature changes are most apparent for ectothermic ("cold-blooded") animals such as insects, reptiles, and fish, for which body temperature, the key determinant of their metabolic rate, strongly tracks the environmental temperature. At lower environmental temperatures, disruption in the availability of energy influences a wide array of physiological and behavioral traits, such as activity patterns and rates of growth and reproduction. In a warming Great Lakes region, we expect that ectotherms like snakes will have longer active periods (prior to becoming dormant for the winter), but higher temperatures, especially in the tropics and deserts, may put these species at high risk if they cannot use behavioral methods to keep cool (Kearney et al. 2009). Homeothermic ("warm-blooded") animals—birds and mammals—maintain a relatively constant body temperature, but still can experience heat-related stress as temperatures continue to increase, especially when they inhabit areas where they are already close to thermal tolerance limits.

Before we give examples of the five types of changes we list above, we want to emphasize that not all changes in species characteristics (phenotypes) require a change at the genetic level. Individuals of many species are able to show flexible responses to temperature as conditions vary among years. Thus, when conditions change in a given location, we can expect to see both "flexible" changes in some species (phenotypic plasticity), and heritable changes (evolution—a change in how common given genes are within the population). In general, phenotypic plasticity can be thought of as a "short-term" solution, as the limits to these responses will eventually be exceeded as a population experiences a long-term increase or decrease in an environmental factor (Gienapp et al. 2008). The potential for evolution in response to climate change is constrained by the degree to which genetic variation for particular traits is present in a given population (Holt 1990). For example, traits that contribute to increased heat or drought tolerance must be present in a population for natural selection to favor the individuals that have those traits, and eventually lead to an overall change in the proportion of individuals that have that "adaptive" trait in later generations. For many of the Great Lakes region's species of greatest conservation concern, we already suspect that population declines, habitat fragmentation, and other stressors have reduced the level of genetic variation such that there is little variation left upon which natural selection can act; however, it is exceedingly rare to actually have data on genetics over time that can be used to confirm or refute this suspicion. Similarly, evidence for genetic responses to climate change is extremely rare, as it requires genetic data to have been sampled over time (Balanyá et al. 2006; see the Genetic Change section below for a few examples). Further, for many species, including some that are able to show flexible responses within a limited range of temperature increases, genetic changes are likely to occur too slowly for natural selection to keep pace with the rapid warming in the environment. As species "fall behind" in terms of adapting to changing conditions, we are highly likely to see more examples of reductions in fitness, population declines, and eventual extinctions. In addition, species that are able to adapt quickly to new conditions may put additional pressures (e.g., as competitors, predators, or parasites) on those that lack sufficient genetic variation, further accelerating the process of species loss.

Shifts in Range Boundaries and Abundance Patterns

The first two types of observed changes on our list, shifts in range boundaries and changes in abundance patterns, both result from changes in the same key "vital rates" (i.e., birth and death rates), but differ in terms of whether we are focused on larger scale (species' ranges) or smaller scale (local abundance) impacts. For many species, changes in climate conditions will enhance a given species' survival rate, growth rate, and/or reproductive rate in some parts of the species' range, and reduce one or more of these rates in other locations. Thus, even without dispersal (movement away from previously occupied habitats), these changes can lead to shifts in the subset of areas within a range where species are common, rare, or absent, and eventual changes in range. Changes in vital rates like survival can be linked back to the physiological constraints of balancing energy reserves under specific climatic conditions, as individuals in highly suitable climatic conditions will often have higher reproduction, survival, or both, than individuals in habitats that are more "costly" (e.g., higher cost of foraging due to heat or cold stress, higher metabolic rate due to higher water temperature for aquatic species).

For a long-term shift into new areas to occur, there must be a way for species to move, a path for them to follow, and a place to go that has climatic conditions that will permit individuals to survive and reproduce. As a general rule, range shifts in response to warming temperatures result in species moving to higher latitudes or altitudes. Movements in mobile species can be direct responses to temperature, such as fish seeking out deeper, colder water, or can be the result of natural selection acting on more random movements by populations of individuals, as those that become established in areas with more suitable climates are more likely to survive and reproduce. Similarly, for species like plants, which are rooted in one location, shifts in range occur as a result of a life stage like seeds being dispersed (e.g., by wind or birds) and becoming established in new areas that are now presumably more suitable than they had been in the past. As much of the Great Lakes region is fairly flat, shifting upwards in altitude is typically not an option; thus, it is likely that terrestrial species would have to migrate for long distances (hundreds of kilometers) over relatively short time periods (several decades) to "follow" their preferred temperature regime. As a result, species that can move rapidly (e.g., birds) are typically seen as more likely to be able to keep up with climate change than other species with lower dispersal capacities (e.g., amphibians, most plants), although of course birds and other mobile species depend on many plants and insects for food and shelter. In addition to moving north within river systems or large lakes, as noted above, some aquatic species also may be able to move into deeper, cooler waters within the same water body, although these deeper habitats may not have all of the other resources that a given species requires.

Examples of species showing range and abundance changes in the Great Lakes region are beginning to accumulate, with the best documented examples coming from researchers conducting long-term research on topics such as community composition and population dynamics. Work by Myers and colleagues (2009) on mammals in Michigan documents rapid changes in ranges for several common species, including northern range edge shifts of over 225 kilometers since 1980 for white-footed mice (*Peromyscus leucopus*) and southern flying squirrels (*Glaucomys volans*). The movement of white-footed mice is of concern from a public health perspective, as these mice are key hosts for the ticks that carry Lyme disease (Ostfeld 1997). The work of Myers et al. (2009) also provides an example that many throughout our region may have recognized in our own lifetimes, and we would not even have had to leave our cars to see the evidence of change. Using distribution data from roadkill surveys conducted in two time periods, 1968 (collected by the Michigan Department of Natural Resources and published in Brocke 1970) and 2006–2008, they document a rapid northern range expansion of our region's only marsupial

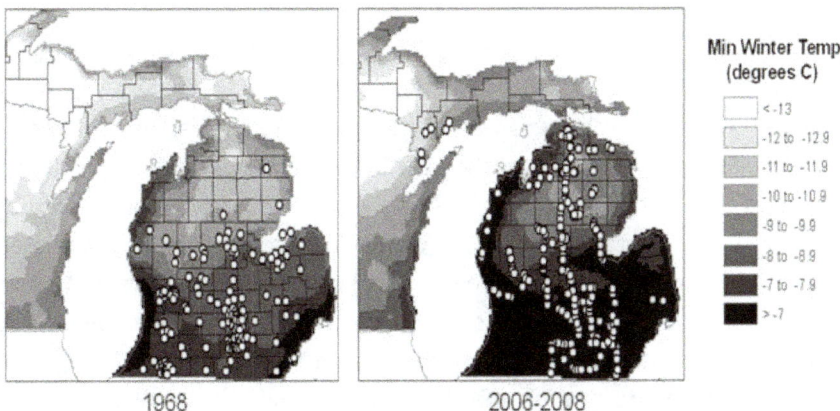

Figure 9.1 Observations of road-killed common opossum from 1968 and 2006–2008, from Myers et al. (2009, figure 4), overlaid upon 20-year average values of minimum winter (December–February) temperature from the period prior to the survey dates (1949–1968, and 1987–2006). The 1968 opossum data were reported by Michigan Department of Natural Resources personnel and presented in Brocke (1970), and were digitized from this source by Myers et al. (2009). The 2006–2008 data were collected by Myers and colleagues, who note that while much of the Upper Peninsula was surveyed, the southwestern corner and eastern edge of the state were not covered. The temperature data are from the PRISM data set at 4 km resolution (Gibson et al. 2002) as available from ClimateWizard (2009).

mammal, the common opossum (*Didelphis virginiana*). In Figure 9.1, we have matched the two data sets from Myers et al.'s work with maps of changes in winter (December–February) minimum temperatures, as this aspect of temperature is likely to have an important influence on where opossums can survive. Note that now, lakes Michigan and Huron seem to form a barrier for further movement into the eastern Upper Peninsula, but opossums seem to be shifting into the western Upper Peninsula from Wisconsin. These three examples highlight species moving in, but Myers et al.'s work also documents reduced ranges for more northern species, and suggests that at least at some well-studied sites, species from the south are replacing those that were formerly dominant, rather than adding to species diversity.

Species are also showing changes in abundance within current ranges. Studies on moose (*Alces alces andersoni*) provide an indication of the complexity of the sensitive relationship between a species' population numbers and environmental temperature. Two separate research teams focused on understanding factors such as birth rates, parasite loads, and survival of moose in northwestern Minnesota (Murray et al. 2006), and on Isle Royale (Vucetich and Peterson 2004; Wilmers et al. 2006). They suggest that warming temperatures are contributing to local population declines. The Minnesota team found that the population growth rate over 1961 to 2000 was negatively associated with higher temperatures, and both teams suggest that warmer temperatures stress moose in a wide variety of ways, from increasing the rate of heat-stress deaths and the impact of parasite infections, to decreasing feeding and birth rates. Both study locations are in the southern part of the species' range, and modeling by the Minnesota team suggests that, given the observed relationships between vital rates and temperature, the northwest Minnesota population of moose will not persist over the next 50 years (Murray et al. 2006). In related work on the dynamics of host-parasite systems in Arctic populations of caribou and musk ox, Kutz et al. (2005) found that one type of nematode (worm-like) parasite can now complete their life cycles in a single year, rather than the three or four years required for reproduction in colder climates. This change has led to rapid increases in the abundance of one species (the parasite) at the expense of the survival of another (the mammalian host; Kutz et al. 2005). Ecologists that specialize in disease dynamics caution that increases in the impacts of current diseases and pathogens are only one climate-change-linked threat to the Great Lakes region's iconic large mammals: as the ranges of other mammals shift farther north, they will likely carry with them diseases and parasites to which current residents of these habitats have little or no natural resistance (Kutz et al. 2005; Dobson 2009).

Changes in species' range boundaries and abundance patterns are of concern for several reasons. First, the rapid changes in climate are taking place in the context of a wide range of other impacts on natural systems, most notably habitat loss and habitat fragmentation due to conversion of natural areas into farms, cities, ports, and other land uses. Further, we expect the impact of humans on the landscape and shoreline to increase as we shift the location of human-changed areas in response to changes in drought risk or lake level, or the need to use more land to support greener energy sources (e.g., wind and solar energy, production of crops and woody biomass for use as biofuels). Highly mobile species like birds and mammals are more likely than many others to be able to shift their ranges fairly rapidly; however, these movements are frequently slowed or blocked by inhospitable landscapes. In the Great Lakes, barriers for terrestrial species can range from the lakes themselves (recall the lack of opossums in the eastern Upper Peninsula of Michigan in Figure 9.1) to large expanses of agricultural fields or pavement. Similarly, dispersal by aquatic species can be limited by the geography of the water body they inhabit, and by human-made barriers such as dams. Ironically, in the Great Lakes region, some of the aquatic barriers have been put into place to protect aquatic systems from damaging invasive species like sea lamprey (*Petromyzon marinus*), so we cannot simply remove them to help species move to more suitable habitats. Even in areas where we have large expanses of intact ecosystems, increasing temperatures can make habitats that depend on particular levels of surface or groundwater more fragmented. Specifically, as evaporation rates increase due to warmer temperatures, even large expanses of coastal wetlands, inland marshes, or wet prairies can become less connected for species like frogs or ducks as some sections dry out, and the remaining wet areas upon which many species depend become more isolated, or disappear.

Second, range and abundance changes are of concern because species that are not able to disperse will have the added stress of species from lower latitudes and altitudes invading their habitats. So, individuals at the southern end of their species' range have the potential to be stressed both by climatic conditions that are becoming less and less favorable and by species that move in from warmer areas and are less challenged by the same climatic factors. The species moving in may directly compete for key resources, and also may contribute to the decline of resident species by spreading diseases and parasites. In the Minnesota moose example described above, the authors suggest that temperature impacts on moose are increased in areas where a more southern species, white-tailed deer (*Odocoileus virginianus*), are abundant, as the deer act as a reservoir for parasites (Murray et al. 2006). In addition to species that act as stressors on wild species of conservation concern, range shifts by species that act as forest or crop pests, or that are detrimental to public health (i.e., carry diseases, create toxic algal blooms) are key concerns in the Great Lakes region, and are important subjects of observational and model-based research studies.

Third, we are concerned about range and abundance shifts because species movements will often be independent of shifts of other species. We expect species to shift independently, as the set of constraints that describe the habitat and ecological niche for each species (factors like temperature, food availability, soil types, and stream flow characteristics) is unique. In effect, we expect to see the "tearing apart" of sets of species that typically interact, and many of these inter-actions may be critical to the survival of one or more of the interacting species. We do have some examples of "specialist" butterfly species increasing the variety of habitats that they use as they shift north (Thomas et al. 2001), but in general, we consider strong dependence on one or a few species to be a strong indicator of vulnerability to climate change. To give an example from the Great Lakes region, the federally endangered Kirtland's warbler (*Setophaga kirtlandii*), which breeds almost exclusively in northern Michigan, is at risk due to a strong dependence on young jack pine trees (*Pinus banksiana*). Jack pine trees grow on sandy, nutrient-poor soils, and are tolerant of cold spring temperatures that can often damage or kill other tree species. For the past three decades, the warbler has shown a dramatic population increase as more and more habitat has been made available, primarily as the result of an intensive collaborative effort between state, federal, and private landowners in Michigan to create large expanses of young jack pine (Donner et al. 2008). However, maintaining jack pine habitat is likely to get more and more challenging as a result of increased drought stress in the summer, increased competition from other trees that are favored by the warmer winters and springs, and increases in the abundance of insect pests formerly limited by cold weather. If our goal is to keep providing habitat for the warbler over the long term, we need to be thinking about how to help facilitate a shift in range, given that the proper soils are relatively rare, and about how to modify management strategies in places that cannot be replaced.

Changes in Phenology

In many organisms, seasonal changes in temperature act as cues that trigger transitions in the species' seasonal cycle, such as metamorphosis (e.g., the transition from egg to larvae), or development of new leaves. The dominant cue for some seasonal changes, like the start date for migration for many birds, is a change in day length (Berthold 1996), but temperature can still have a strong influence on the timing of migration by influencing the rate at which birds travel from their wintering grounds to breeding habitats. In addition to directly triggering changes in timing, warming trends can impact species by influencing other key seasonal events that trigger changes in their seasonal cycles, such as timing of snowmelt or flooding, or lake stratification. The term "phenology" describes the timing of these seasonal events, and recording the date of events like bird arrivals, tree fruiting, or ice breakup on lakes has been of interest to both scientists and members of the general public for decades—and even centuries in some locations. As a result of this broad interest in the timing of seasonal events, the scientific community has had access to long-term data sets showing strong, easy-to-observe responses by species to changes in temperature. These long-term data sets have played a key role in the evaluation of climate change impacts, and in raising public awareness about the risks to biodiversity posed by what many might think of as "inconsequential" changes in temperature.

The strength of phenological data as a tool for raising awareness of the impacts of climate change is hard to overemphasize. When aggregated across large scales, these data show very strong patterns, and because many people are familiar with the types of events (e.g., timing of bird arrivals or leaf opening in spring) that are being measured, changes can be clearly communicated to nonscientists. Assessing broad patterns in responses to climate change is possible with phenology data because various types of changes seen in a wide range of species can be measured in the same unit: number of days earlier or later. This ability to group observations of change across species and continents using the same measure of change has facilitated broad "meta-analyses" that link shifts in timing to local changes in temperature (e.g., Root et al. 2003, 2005; Parmesan and Yohe 2003), and has facilitated a rapid increase in the recognition of climate change as a threat to biodiversity. Several early phenology studies that were highly influential in raising awareness that species were responding to changes in climate focused on, or included, study sites in the Great Lakes region. These included evidence of 10- to 13-day advances in frog-calling dates (an indicator of breeding) in western New York in response to a 1° to 2.3°C increase in temperature in key months (Gibbs and Breish 2001), advances in the timing of many spring events (bird arrivals, plant blooming) on a Wisconsin farm in the 1980s and 1990s relative to observations taken by Aldo Leopold in the 1930s and 1940s (Bradley et al. 1999), and a nine-day advance in the laying date of tree swallows (*Tachycineta bicolor*) across the continental United States over 32 years (1959–1991; Dunn and Winkler 1999).

Overall, the patterns of change documented in phenological studies provide strong evidence for a "fingerprint" of climate change on the timing of seasonal events. In the simplest form of comparison, phenological responses to increasing temperature can be grouped into three patterns: shifts toward earlier phenology, shifts later, and no change (Root et al. 2003; Parmesan and Yohe 2003; Root et al. 2005; Parmesan 2006). Focusing on the first two groups (shifts earlier and later), we would expect to see a roughly equal (or random) number of species exhibiting each type of change if increasing temperature was not a causal factor. Instead, in a comparison by Root et al. (2003) of published records of timing of spring events from 700 species exhibiting statistically significant change over the last 30 years in locations from across the globe, only 6 out of the 700 species (<1%) showed a shift toward later timing. Overall, these data suggest a strong pattern of spring events occurring earlier across different types of species and around the world, at a rate of around 5 days per decade (Root et al. 2003).

In addition to helping us understand the direction and rate of change that species are exhibiting, phenological data have also played an important role in linking shifts in species to human actions that contribute to increasing CO_2 like burning fossil fuels and deforestation. Root et al. (2005) compared temperature predictions from global circulation models (GCMs) derived from model runs using only natural forcings (e.g., volcanic dust), only human forcings (e.g., CO_2 emissions), or a combination of both natural and human forcings. By comparing the observed changes in species with these differently modeled temperature predictions, they were able to see the influence of natural and anthropogenic drivers of change. When only natural forcings were used, the associations between the predicted temperatures and observed changes in species were quite weak, while with only anthropogenic forcings, the correlations were stronger, and with a combination of both forcings, the fit between modeled and observed changes was quite strong (Root et al. 2005).

Quantifying the overall percentage of species that are changing is harder than comparing the proportions showing shifts earlier or later. This is because we know that some species are not exhibiting changes in the timing of their spring events, even in locations where temperatures have increased over time. However, when we try to capture global trends in a review of literature published over many decades, we run into the problem of publication bias. Specifically, it is much less likely that research documenting "no change," or an insignificant relationship between temperature and a change in timing, will be published unless this information is reported along with reports of significant change in another species. There are several reasons why changes may not be detected; some changes may be triggered by other factors like day length, or seasonal weather may be highly variable, such that it takes several decades or longer than the duration of a given study to detect a statistically significant trend. However, lack of a phenological response does not mean that species are "immune" to temperature changes, for at least two reasons. First, they may be responding in other ways not addressed in the same study, for example by changing behavior or distribution, and second, a lack of response may have negative consequences (see "Phenology Mismatches" discussion, below). Multi-species monitoring networks provide essential data sets for helping us tease apart what factors determine which species respond, and the consequences of responding or not. Thus, continued support of new and existing networks of monitoring is an important investment that will help us determine what actions to take to better protect species over the long term.

Although examples abound of changes in phenology, indications of what these changes mean in terms of risk to species from climate change requires comprehensive data sets that go beyond just measures of timing. A recent paper documenting long-term (approximately 100 years) changes in phenology and abundance of 429 plant species in Concord, Massachusetts (many of which are also found in the Great Lakes region), showed that although there has been an overall shift of 7 days in flowering phenology associated with a 2.4°C temperature increase in the study area, some plant families are showing less of a response to temperature than others (Willis et al. 2008). In many cases, this failure to shift flowering time in response to changes in seasonal temperature was associated with strong declines in abundance (Willis et al. 2008). Thus, rather than just documenting a change in timing, this paper represents a major contribution to helping us understand why we should be concerned about phenological changes (or lack of change), as it links the timing change to a change in fitness through the measure of abundance. Of particular interest, Willis and colleagues (2008) found that the plant species showing declines were more closely related than expected by chance. As research continues on both genetic and flexible (phenotypic) responses to climate change, it seems likely that broad patterns will continue to be identified that help us to evaluate the responses of species by group, rather than trying to plan for the idiosyncratic responses of each species.

Phenology Mismatches

Rapid phenological changes are of concern because for tens of thousands of years or more, wild species have been adjusting to seasonal changes in their environment, and to associated changes in species that act as predators, prey, parasites, and competitors. By contributing to the buildup of greenhouse gases, we have dramatically increased the rate of change, and more and more studies are suggesting that species in the same area are not responding at the same rate. Our fourth category of observed changes, phenology mismatches, describes situations where species that interact in some important way respond differently to a temperature change. The potential importance of mismatches may be easiest to imagine in systems where attainment of a threshold temperature cues the emergence of leaves of a dominant tree or grass, or algal growth. In such a system, a shift in the timing of spring warming that alters when these plants grow or bloom could represent a key change in the foundation of the food web that determines energy flows throughout that entire ecological system. If other species in the same system do not shift in the same direction and at a similar rate, they may be at a strong disadvantage in terms of their ability to survive and reproduce relative to other species with similar resource requirements. Evidence of the potential for cascading phenology mismatches in lake food webs comes from work by Winder and Schindler (2004) in Washington State. In their focal lake, they found that phytoplankton (algae) have been blooming earlier, with some types of zooplankton (tiny animals that eat algae) showing similar changes in timing, and others not, potentially impacting food availability for a wide range of fish and other predators (Winder and Schindler 2004). As noted above, conditions in the Great Lakes are changing rapidly (increasing temperature, longer stratified period, stronger currents), suggesting a high potential for species to respond at different rates and contribute to disruption of entire food webs.

Although a wide variety of species are likely vulnerable to phenological mismatches, we are not aware of any research that has focused on this topic within the Great Lakes region. However, it is not very hard to pull together information that makes a compelling argument that these types of changes should be of concern, and we do this here using the example of songbirds that migrate through, and breed in, the Great Lakes region. The northern Great Lakes region stands out within North America for supporting a high diversity of breeding songbird species (Price et al. 1995), and the region also supports vast numbers of birds during spring and fall migration. Most songbirds depend upon a ready source of insect prey, both along their migration routes and in their breeding habitats. Studies in Europe have documented advances in insect emergence relative to bird arrivals at breeding habitats, and suggest that these timing mismatches are leading to reduced breeding success (Both and Visser 2001; Visser et al. 2006).

To assess the vulnerability of migratory birds to changes in plant or insect phenology, we need a "yardstick" for understanding the impact of a particular magnitude of change (or lack of change) in phenology (Visser and Both 2005; Visser 2008; Both et al. 2009). The work cited above, along with work by Marra et al. (2005), suggests that changes in bird migration phenology may be slower than the responses of many of the plants and insects at the stopover sites upon which these birds depend. Marra et al. (2005) compared the median capture dates of 15 long distance migrants from bird monitoring stations in coastal Louisiana and two stations in the Great Lakes region, Long Point Bird Observatory (on the north shore of Lake Erie) and Powdermill (western Pennsylvania). They also compared the duration of time between the median arrivals for the same species at the southern and northern sites. Marra et al. (2005) found that median capture dates were earlier in years with warmer spring temperatures (mean April/May temperature) for almost all of their focal species, at a rate of roughly 1 day earlier per each 1°C increase in temperature. However, they note that lilac (*Syringa vulgaris*) budburst occurred 3 days earlier for the same temperature increment, a similar rate to the average reported for plants in the Willis et al. (2008) study described above. Similarly, Strode (2003) suggests that North American wood warblers are not advancing in phenology as fast as key prey are likely to be responding to increased temperatures (e.g., the eastern spruce budworm, *Choristoneura fumiferana*). Marra et al.'s results of earlier median passage in warm springs is still consistent with our earlier statement that a key cue for migration departure is day length; earlier arrivals were at least in part achieved through faster migration (as opposed to earlier departure dates) as the duration of migration between the southern and northern locations decreased by 0.8 days with every 1°C increase (average of 22 days; Marra et al. 2005).

In the Great Lakes region, migratory songbirds may be most at risk by being at the right place at the wrong time at migratory "stopover" habitats at the edges of the upper Great Lakes, due to rapid warming of lake surface waters relative to nearby land (described above, Austin and Colman 2007, 2008; Dobiesz and Lester 2009). This pattern suggests that even if birds manage to track the rate of phenological shifts occurring on land near the shoreline, they may be too late to take advantage of the emergence of aquatic insects upon which many birds depend while refueling during migration (Ewert and Hamas 1995; Smith et al. 1998). If this occurs, the timing mismatch could cascade through the migratory process, with birds having to stay longer at the stopover site in order to regain energy reserves for sustained flying (Moore et al. 2005).

Even without considering whether coastal stopover sites are warming faster than inland sites, changes in phenology along migration routes are likely to increase the risk of mortality, or at least the potential for reduced fitness, of migrating birds. This is because migratory birds are likely to encounter strong geographic variation in the extent to which temperature trends, and as a result plant and insect phenology, are changing: Some sites show strong warming trends, while others may have warmed less over time, or even cooled. To illustrate this point, in Figure 9.2 we have drawn two hypothetical migration routes for birds breeding in the northern Great Lakes region, and show them over a map of trends in spring maximum temperature for a 50-year period

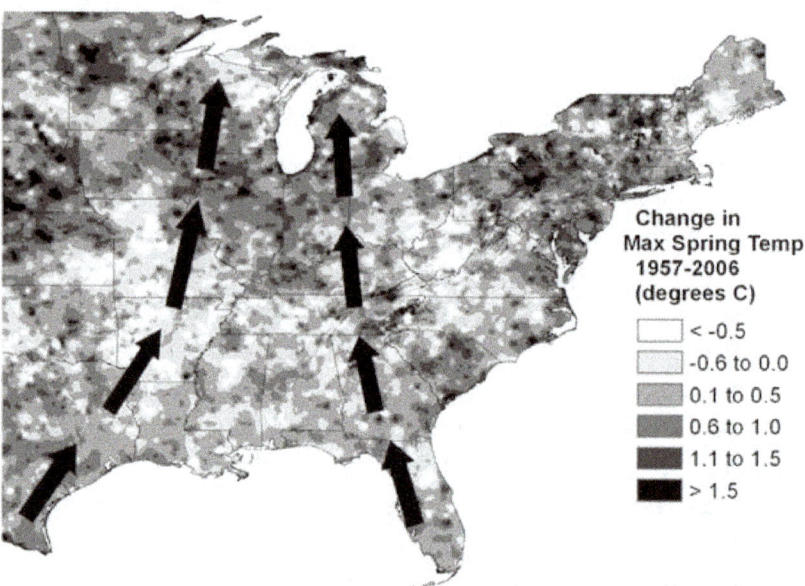

Figure 9.2   Variation in estimates for net change in the average spring (March–May) maximum temperature over a 50-year period (1957–2006). The arrows indicate two hypothetical pathways that birds might follow as they migrate through the southern U.S. toward breeding grounds in the northern Great Lakes region. Note that although much of our region is showing peak temperature increases, birds will have to move through regions where peak spring temperatures are showing decreasing trends, suggesting birds may encounter high variation in phenology of key plants and insect prey along their migratory routes. The temperature data are from the PRISM data set at 4 km resolution (Gibson et al. 2002), with 50-year trend calculated by ClimateWizard (2009; see Girvetz et al. 2009 for details).

(1957 to 2006). Note that although the central United States is experiencing an overall annual warming trend, maximum spring temperature, a factor that is likely highly correlated with plant and insect phenology, has been showing both increases (dark areas) in some areas, and even slight decreases (white to light gray) in other areas during this time period. So if birds were to follow the arrows in the diagram, though they would encounter a lot of local variation, in general they would first encounter areas that are showing moderate warming (e.g., Texas, Florida), then reach areas that are warming less or even cooling (Arkansas, much of Tennessee), and finally reach areas, potentially including their breeding grounds, that are warming much more rapidly (much of Minnesota, western Wisconsin, and Michigan). Thus, while species that migrate are obviously highly mobile and able to rapidly shift their distribution, the potential for reduced synchrony with key resources along migration routes suggests that they may be particularly vulnerable to the impacts of climate change.

[...]

Changes in Precipitation

While our review focuses on the biological theory and observations related to species responses to changes in air or water temperature, many other aspects of species' environments are also changing due to global warming. These include factors like precipitation patterns, storm intensity, drought stress, ice cover, and lake level, all of which may interact with, or even counteract, some effects of temperature increase. When compared to temperature increases, changes in the patterns of precipitation are harder to generalize into "rules of thumb" that can help us anticipate species' vulnerability. For average precipitation in particular, predictions from different GCMs (there are about 16 in common use, as well as many regional models) for the Great Lakes region are highly variable across space and time, and there is more uncertainty in both the direction of magnitude of changes projected and the time of year in which any projected changes will likely occur. Thus, when thinking about precipitation in particular, probably the most important rule is that we must be prepared to be more adaptive in our conservation and management actions, and to inform these actions using the range of possible outcomes suggested by modeling work. In other words, we need to think both about how species may respond to more rain in various seasons, and to less, and to remember also that even if the amount of rain or snow stays about the same, it will evaporate or melt faster in a warmer climate.

The same "bet hedging" approach should be applied to planning for changes in water levels of the Great Lakes, as these are influenced by air temperature regime (which drives losses of water through evaporation), water temperature regime (including the feedback between ice cover and water temperature described above), and by precipitation (for more detail on lake-level processes, see Mackey, this volume). While projections for future lake levels that focus on mean values suggest steady declines in lake levels over time (e.g., Hayhoe et al. 2010a, also cited in CCSP 2009), work by Angel and Kunkel (2010) demonstrates that when you explicitly consider the variation across GCMs and across emissions scenarios, lake-level models suggest a wide range of potential changes, including slight increases. This variation primarily results from differences across GCMs in projections for the amount and timing of precipitation, which is a persistent challenge for climate modelers, and one that is not likely to be resolved soon. Thus, the most pragmatic approach is to develop conservation plans that anticipate drops in mean lake levels, but that also benefits species if the seasonal range of lake levels stays about the same or even increases. Further, we need to also be thinking about how human use of these precious water resources might change, and how to include protection of ecosystems and species into decisions that are made regarding future water-use regulation policies.

With respect to extreme precipitation events rather than mean values, however, there is general agreement that the frequency of extreme rain events (intense storms) will increase, especially in the winter and spring. Trends over the last 50 years for the upper Midwest suggest about a 30 percent increase in the amount of rain that falls in the top 1 percent of "very heavy" precipitation events, and this impact is expected to increase due to the fact that warmer air can hold more water (CCSP 2009, based on updates to Groisman et al. 2004). The impacts of this change on aquatic systems can be quite strong, especially in landscapes with high proportions of agricultural or urban land uses, which act as sources of pollutants and fertilizers when large volumes of water flow across them into rivers and coastal areas. Similarly, we expect increases at the other end of the extreme weather events spectrum, in the form of summer droughts.

While precipitation changes are hard to predict, there are many things that we can do to safeguard the species and natural systems of the Great Lakes through these ongoing changes. We have this opportunity to do a better job protecting species because many of our land-use and infrastructure decisions directly influence how water moves through our landscapes, and many of the key stressors on biodiversity, especially in aquatic systems, result from impacts that are highly influenced by extreme rain events. As noted above, much of the sediment and pollution that enters rivers and nearshore systems does so during extreme storm events that cause large volumes of water to flow across agricultural fields and cities, and contributes to overflows of combined sewage and stormwater handling facilities. Actions such as restoring wetlands that can absorb water, promoting "best management practices" on farms, and improving our sewage handling and other pollution-prevention infrastructure can all benefit wild species by minimizing pollution and sedimentation from storm events. Further, when actions like wetland restoration, and restoration of functional floodplains, are implemented to help handle stormwater, species also reap the benefit of increased habitat area and enhanced connectivity between habitat patches. These kinds of "green infrastructure" investments also benefit people by reducing the risk of flooding, and by helping to promote clean water for drinking and recreation, and in many cases may provide lower-cost, more reliable strategies for containing flow than traditional infrastructure-based options.

Understanding and Responding to Species Vulnerabilities

Thus far, the weight of evidence suggests that the most appropriate expectation for how species may respond to climate change is to anticipate more of the types of changes we have already seen—i.e., changes in ranges (evading the temperature change), and changes in phenology and behavior that allow species to persist in the same range. We need to remember, however, that many of these "stay in place" changes may be the result of phenotypic plasticity, suggesting that in the coming decades, many species that appear to be adapting at a rate that allows them to track changes in climate may show sudden declines in viability once the temperature shift exceeds some critical threshold beyond which their "flexible" response is not enough. As of yet, while there are many examples of changes in species in response to climate change, there are no documented examples of genetic shifts in thermal tolerances that appear to allow species to remain viable in the same location following a change that would have otherwise led to reduced survival or reproduction (Parmesan 2006; Bradshaw and Holzapfel 2008; Gienapp et al. 2008). Even given these caveats, developing our understanding of how we could modify our management and conservation priorities to account for shifts in ranges (i.e., promoting connectivity between habitats) and in phenology (ensuring protection of species across a range of current and future microclimates) is an important place to start making our ongoing work "climate smart."

Based on the wide range of observed impacts described here, it is easy to see that predicting impacts on, and responses by, a given species or system can get complicated very quickly. To illustrate this complexity, we have compiled a list of climate factors that are changing in the Great Lakes region, and examples of how these can impact species (Figure 9.3). Many of these impacts, such as changes in temperature, have broad-reaching impacts, while others are likely to impact relatively small subsets of species through very specific mechanisms. For example, the figure notes that increases in CO_2, the major driver of changes in temperature, can also influence competitive relationships between species. This impact applies specifically to plants, and acts by changing the relative efficiency of different variations on the process of photosynthesis, the mechanism by which plants convert energy from the sun into new plant tissues. Because the suite of potential impacts is so large, and impacts are often interrelated, our best guesses on impacts and species vulnerability may vary considerably depending on how many of these factors are considered. For example, Jones et al. (2006) found that projections of the potential impact of climate change on Lake Erie walleye (*Sander vitreum*) based simply on water-temperature change were very different from results incorporating changes in climate-sensitive factors such as water levels and light penetration. This work relied upon decades of research on this fish's habitat needs and biology, and illustrates that for well-known species like walleye, the challenge to managers and conservation practitioners may focus on characterizing a complex set of direct and indirect climate-related changes that may interact and influence species survival. For most other species, a lack of baseline information from which to even begin the process of understanding potential impacts is often the most daunting challenge.

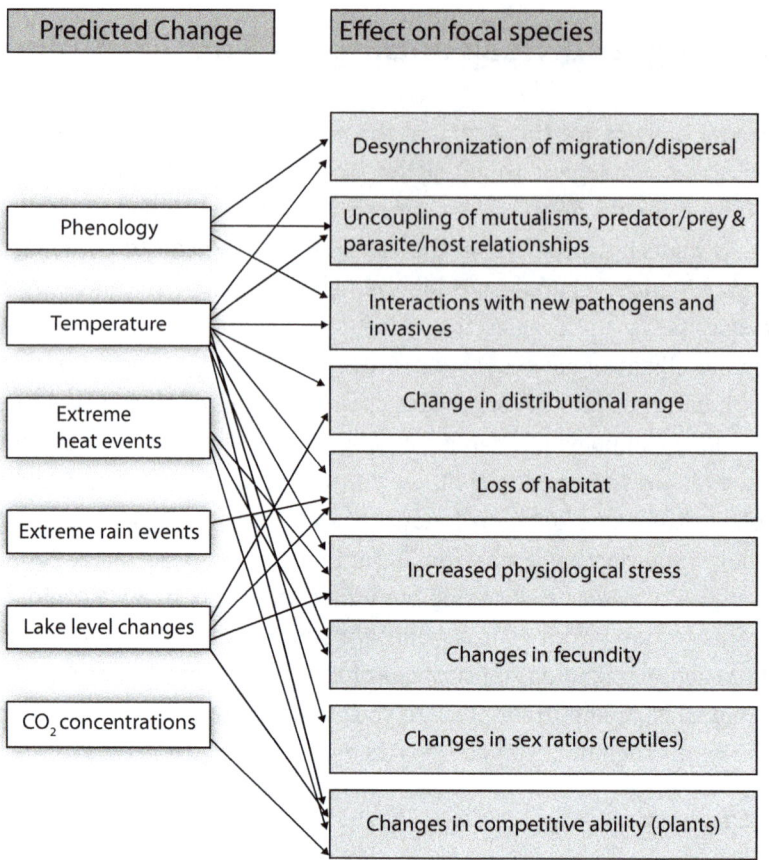

Figure 9.3 Illustration of the many ways in which climate change can impact species in the Great Lakes region, adapted from Foden et al. (2008, fig. 1). Note that the climate factors and effects are not mutually exclusive—as, for example, the driver "phenology" represents a change in the timing of some factor like temperature or precipitation.

Finally, we will conclude this chapter with some general rules of thumb for identifying highly vulnerable species, and examples of the kinds of "surprises" that will require us to be much more agile in how we plan and implement actions to benefit wild species. Characteristics often identified as indicators of species that are at greatest risk of population decline, or possibly even extinction due to climate change impacts, include:

- Occurrence at high altitude or latitude (cannot shift range farther up or north)
- Occurrence in isolated habitats surrounded by inhospitable land, either developed or undeveloped, that inhibits dispersal
- Being near limits of physiological tolerance
- Having limited dispersal and/or colonizing ability

- Having very specific habitat requirements
- Being highly dependent on interactions with one or a few other species (susceptible to phenology mismatches, and mismatches in rate or location of range shifts)
- Having long generation time (slow potential pace of microevolution)
- Having low genetic variability and/or low phenotypic plasticity.

Although applying these and other much more comprehensive and site-specific approaches to assessing vulnerability is a major challenge, an even greater challenge is likely to be deciding on what percentage of our limited resources should go towards trying to protect those species identified as most vulnerable. Given the projections for rapid increases in species extinctions, we need to think carefully about how to make decisions that protect the widest range of species and functional ecosystems, some of which the public may tend to value, and others that may be less well known or appreciated, but are still critical to the long-term maintenance of biodiversity.

As we work to update our conservation plans and make them "climate smart," it is vitally important that we also update our approaches to management such that they become more agile and able to shift strategies quickly in the face of new information and surprises. Acting in a climate-smart way will also require that we improve our ability to share and synthesize the information we do have, and improve our tools for acting in the face of uncertainty, as discussed in detail in the chapters by Scheraga, Moser, Marx and Weber, and Easterling et al. in this volume. With respect to anticipating surprises, we expect that surprises for resource managers will take at least three forms: (1) exceedance of thresholds (e.g., thermal tolerance thresholds, leading to strong declines in fitness); (2) new interactions among species, and/or new or synergistic impacts related to interactions with climate and other stressors (e.g., invasive species); and (3) higher frequency of extreme weather events with catastrophic impacts on focal systems (floods, ice storms, extreme cold periods in spring).

In some cases, surprises may result from a combination of factors and threshold exceedance, and be surprising due to a mix of positive and negative impacts on fitness. For example, work by Tucker et al. (2008) shows that many red-eared sliders (aquatic turtles, *Trachemys scripta elegans*) in Illinois have been laying an additional clutch (three instead of two) of eggs. The increase in reproduction has been achieved by turtles extending their breeding season 1.2 days per year from 1995 to 2006, and initiating breeding earlier at a rate of 2.2 days per year over the same time period, which showed significant spring to summer warming. However, like other turtles, the sex ratio for this species is temperature dependent, and these later clutches are being deposited into below-ground nests when soil temperatures are low relative to a temperature threshold that promotes an even ratio, leading to a male bias in the offspring. This male bias is surprising, as the general prediction based on warming would be for these turtles to produce more females. However, with turtles nesting earlier and later (first and third clutches), the authors suggest that only the second clutches would be likely to produce many females. Clearly, if the population becomes strongly male-biased, this would suggest population declines.

Some serious risks to public health and wildlife habitat may be in store for us as a result of one species of invasive cyanobacterium, or blue-green algae, that embodies two of the above forms of surprises: thresholds and new invasive species. Recent work by Hong et al. (2006) documents that *Cylindrospermopsis raciborskii*, a toxin-producing and bloom-forming species, has been found in Muskegon and Mona lakes in western Michigan, which connect directly to Lake Michigan. First, *C. raciborskii* is an example of a species that seems to require exceedance of a temperature threshold of roughly 22°C (Hong et al. 2006, and work reviewed therein) to germinate from an inactive, climate-resistant spore, to the motile form that creates blooms and produces toxins in the water column. Second, it has "surprise" potential because this invasive species is a good competitor with other phytoplankton, and we don't know how it moves, or where the spore is likely to be waiting in lake sediments for suitable temperatures to arise (Hong et al. 2006).

In our last example of surprises, we focus on a change that many people who do not depend on income from the Great Lakes region's winter-recreation sector might be looking forward to—earlier spring warming. As described above, we have lots of evidence that species respond to changes in the timing of seasonal events like spring warming, which may suggest that timing changes are not likely to act as stressors on populations. However, in addition to potentially contributing to timing mismatches, shifts in timing may lead to surprises by making species more vulnerable to extreme weather events that are typical for a given season. For example, a period of record low temperatures occurred across much of the eastern United States in April of 2007 after what had been an unusually warm later winter/early spring (Gu et al. 2008). The warm spell led to early bud break and canopy leaf-out in trees, and these sensitive plant tissues were then blasted by a very cold arctic air mass in early April, leading to massive damage to plants across a variety of systems, especially in centrally located states like Tennessee and North Carolina. These kinds of events may be a potential threat to species' viability, both because they can lead to direct reductions in survival or fitness when the event occurs and because they may have longer-term impacts on future generations by producing selective pressures in opposition to pressures presented by gradual climatic changes (i.e., increases in mean temperature). For example, an early spring cold snap in Nebraska killed a large proportion of cliff swallows (*Pterochelidon pyrrhonata*) that had arrived early following migration (Brown and Brown 2000). In later years, birds at these same colonies arrived significantly later, providing evidence of directional selection on arrival timing (Brown and Brown 2000). Assuming that local and regional warming would favor birds that arrive earlier in most years, intense selection from these rare weather events could be counteracting an otherwise beneficial response.

In conclusion, we would like to reemphasize that based on species responses to climate change that we can already observe, and on the theories that frame our understanding of these changes, models project high rates of extinction if we continue to rely upon carbon-based fuels. Estimates of range and abundance changes in response to various future scenarios (e.g., a doubling of atmospheric CO_2) suggest that between 15 percent and 40 percent (Field et al. 2007, based on Thomas et al. 2004) of all known species will go extinct due to human enhancement of atmospheric greenhouse gases. Given that there are around 1.75 million species that have been described, somewhere between 250,000 and 700,000 known species, to say nothing about the unknown species, could go extinct primarily due to our use of fossil fuels and the dumping of their combustion products into the atmosphere as if it were an unpriced sewer. As a northern region with many rare species, the Great Lakes region is at risk, and we need to take action now to slow the rate of change and protect the biodiversity upon which we as a society depend. Further, we need to act now to help safeguard species against changes that are already occurring, and which we know are coming due to emissions that have already been released. Protecting species and functional ecosystems will require rethinking our management and conservation strategies to ensure that they are climate smart, and that we become much more efficient and targeted in our investments in natural resources. These investments must go beyond on-the-ground actions to include significant investment in tools for managing and sharing information across the region, and tools to help us improve our ability to make decisions in the face of uncertain information.

References

Angel, J.R., and K.E. Kunkel. 2010 The response of Great Lakes water levels to future climate scenarios with an emphasis on Lake Michigan-Huron. *Journal of Great Lakes Research* 36 (Suppl. 2): 51–58.

Austin, J.A., and S.M. Colman. 2007. Lake Superior summer water temperatures are increasing more rapidly than regional air temperatures: A positive ice-albedo feedback. *Geophysical Research Letters* 34: L06604, doi: 06610.01029/02006GL029021.

——. 2008. A century of warming in Lake Superior. *Limnology and Oceanography* 53:2724–2730.

Balanyá, J., J.M. Oller, R.B. Huey, G.W. Gilchrist, and L. Serra. 2006. Global genetic change tracks global climate warming in *Drosophila subobscura*. *Science* 313:1773–1775.

Berthold, P. 1996. *Control of Bird Migration*. London: Chapman and Hall.

Both, C., and Visser, M.E. 2001. Adjustment to climate change is constrained by arrival date in a long-distance migrant bird. *Nature* 411:296–298.

Both, C., M. van Asch, R.G. Bijlsma, A.B. van den Burg, and M.E. Visser. 2009. Climate change and unequal phenological changes across four trophic levels: Constraints or adaptations? *Journal of Animal Ecology* 78:73–83.

Bradley, N.L., A.C. Leopold, J. Ross, and H. Wellington. 1999. Phenological changes reflect climate change in Wisconsin. *Proceedings of the National Academy of Sciences* 96:9701–9704.

Bradshaw, W.E., S. Fujiyama, and C.M. Holzapfel. 2000. Adaptation to the thermal climate of North America by the pitcher-plant mosquito, *Wyeomyia smithii*. *Ecology* 81:1262–1272.

Bradshaw, W.E., and C.M. Holzapfel. 2008. Genetic response to rapid climate change: It's seasonal timing that matters. *Molecular Ecology* 17:157–166.

Brocke, R. 1970. The winter ecology and bioenergetics of the opossum, *Didelphis marsupialis*, as distributional factors in Michigan. PhD dissertation, Michigan State University, East Lansing, MI.

Brown, C.R. and M.B. Brown. 2000. Weather-mediated natural selection on arrival time in cliff swallows (*Petrochelidon pyrrhonata*). *Behavioral Ecology and Sociobiology* 47:339–345.

Brown, J.H. 1995. *Macroecology.* Chicago: University of Chicago Press.

CCSP. 2009. *Global Climate Change Impacts in the United States.* U.S. Climate Change Science Program, Unified Synthesis Product. U.S. Climate Science Program, Washington, DC. http://www.globalchange. gov/publications/reports/scientific-assessments/us-impacts.

Christensen, J.H., B. Hewitson, A. Busuioc et al. 2007. Regional climate projections. In *Climate Change 2007: The Physical Science Basis*, ed. by S. Solomon, D. Qin, M. Manning et al., 847–940. Cambridge: Cambridge University Press.

ClimateWizard. 2009. ClimateWizard is a web-based tool for making past and projected climate data from various sources more accessible, and was created through a collaboration between The Nature Conservancy, the University of Washington, and the University of Southern Mississippi (http://www. climatewizard.org). The PRISM data shown here were created February 4, 2007, and were processed with ClimateWizard in August 2009.

Desai, A.R., J.A. Austin, V. Bennington, and G.A. McKinley. 2009. Stronger winds over a large lake in response to weakening air-to-lake temperature gradient. *Nature Geoscience*: doi: 10.1038/NGE0693.

Dobiesz, N.E., and N.P. Lester. 2009. Changes in mid-summer water temperature and clarity across the Great Lakes between 1968 and 2002. *Journal of Great Lakes Research* 35:371–384.

Dobson, A. 2009. Climate variability, global change, immunity, and the dynamics of infectious diseases. *Ecology* 90:920–927.

Donner, D.M., J.R. Probst, and C.A. Ribic. 2008. Influence of habitat amount, arrangement, and use on population trend estimates of male Kirtland's warblers. *Landscape Ecology* 23:467–480.

Dunn, P.O., and D.W. Winkler. 1999. Climate change has affected the breeding date of tree swallows throughout North America. *Proceedings of the Royal Society of London Series B–Biological Sciences* 266:2487–2490.

Etges, W.J., and M. Levitan. 2008. Variable evolutionary response to regional climate change in a polymorphic species. *Biological Journal of the Linnean Society* 95:702–718.

Etterson, J.R., and R.G. Shaw. 2001. Constraint to adaptive evolution in response to global warming. *Science* 294:151–154.

Ewert, D.N., and M.J. Hamas. 1995. Ecology of migratory landbirds during migration in the Midwest. In *Management of Midwestern Landscapes for the Conservation of Neotropical Migratory Birds*, ed. by F.R. Thompson III, 200–208. U.S. Dept. Agriculture, Forest Service, North Central Forest Experiment Station, Gen. Tech. Rep. NC-187.

Field, C.B., L.D. Mortsch, M. Brklacich et al. 2007. North America. In *Climate Change 2007: Impacts, Adaptation and Vulnerability*, ed. by M.L. Parry, O.F. Canziani, J.P. Palutikof, P.J. van der Linden, and C.E. Hanson, 617–652. Cambridge: Cambridge University Press.

Foden, W., G. Mace, J.-C. Vié et al. 2008. Species susceptibility to climate change impacts. In *The 2008 Review of the IUCN Red List of Threatened Species*, ed. By J.-C. Vié, C. Hilton-Taylor, and S.N. Stuart. Gland, Switzerland: IUCN.

Gibbs, J.P., and A.R. Breisch. 2001. Climate warming and calling phenology of frogs near Ithaca, New York, 1900–1999. *Conservation Biology* 15:1175–1178.

Gibson, W.P., C. Daly, T. Kittel et al. 2002. Development of a 103-year high-resolution climate data set for the conterminous United States. In *Proceedings, American Meteorological Society, Portland, OR, May 13–16*, 181–183. http:// www.prism.oregonstate.edu/pub/prism/docs/appclim02103yr_hires_dataset-gibson. pdf.

Gienapp, P., C. Teplitsky, J.S. Alho, J.A. Mills, and J. Merilä. 2008. Climate change and evolution: Disentangling environmental and genetic responses. *Molecular Ecology* 17:167–178.

Girvetz, E.H., C. Zganjar, G.T. Raber, E.P. Maurer, P. Kareiva, and J.J. Lawler. 2009. Applied climate-change analysis: The Climate Wizard tool. PLoS ONE 4:e8320. doi: 8310.1371/journal.pone.0008320.

Groisman, P.Y., R.W. Knight, T.R. Karl, D.R. Easterling, B. Sun, and J.H. Lawrimore. 2004. Contemporary changes of the hydrological cycle over the contiguous United States, trends derived from *in situ* observations. *Journal of Hydrometeorology* 5:64–85.

Gu, L., P.J. Hanson, W. Mac Post et al. 2008. The 2007 eastern US spring freezes: Increased cold damage in a warming world? *Bioscience* 58:253–262.

Hayhoe, K., J. VanDorn, T. Croley II, N. Schlegal, and D. Wuebbles. 2010a. Regional climate change projections for Chicago and the US Great Lakes. Journal of Great Lakes Research 36 (Suppl. 2):7–21.

Hayhoe, K., S. Sheridan, L. Kalkstein, and S. Greene. 2010b. Climate change, heat waves, and mortality projections for Chicago. *Journal of Great Lakes Research* 36 (Suppl. 2):65–73.

Holt, R.D. 1990. The microevolutionary consequences of climate change. *Trends in Ecology and Evolution* 5:311–315.

Hong, Y., A. Steinman, B. Biddanda, R. Rediske, and G. Fahnenstiel. 2006. Occurrence of the toxin-producing cyanobacterium *Cylindrospermopsis raciborskii* in Mona and Muskegon Lakes, Michigan. *Journal of Great Lakes Research* 32:645–652.

Jones, M.L., B.J. Shuter, Y. Zhao, and J.D. Stockwell. 2006. Forecasting effects of climate change on Great Lakes fisheries: Models that link habitat supply to population dynamics can help. *Canadian Journal of Fisheries and Aquatic Sciences* 63:457–468.

Kearney, M., R. Shine, and W.P. Porter. 2009. The potential for behavioral thermoregulation to buffer "cold-blooded" animals against climate warming. *Proceedings of the National Academy of Sciences* 106:3835–3840.

Kutz, S.J., E.P. Hoberg, L. Polley, and E.J. Jenkins. 2005. Global warming is changing the dynamics of Arctic host-parasite systems. *Proceedings of the Royal Society B* 272:2571–2576.

Levitan, M. 2003. Climatic factors and increased frequencies of "southern" chromosome forms in natural populations of *Drosophila robusta*. *Evolutionary Ecology Research* 5:597–604.

Magnuson, J.J., K.E. Webster, R.A. Assel et al. 1997. Potential effects of climate changes on aquatic systems: Laurentian Great Lakes and Precambrian Shield Region. *Hydrological Processes* 11:825–871.

Marra, P.P., C.M. Francis, R.S. Mulvihill, and F.R. Moore. 2005. The influence of climate on the timing and rate of spring bird migration. *Oecologia* 142:307–315.

McCormick, M.J., and G.L. Fahnenstiel. 1999. Recent climatic trends in nearshore water temperatures in the St. Lawrence Great Lakes. *Limnology and Oceanography* 44:530–540.

Meehl, G.A., T.F. Stocker, W.D. Collins et al. 2007. Global climate projections. In *Climate Change 2007: The Physical Science Basis*, ed. by S. Solomon, D. Qin, M. Manning et al., 747–843. Cambridge and New York: Cambridge University Press.

Millien, V., S.K. Lyons, L. Olson, F. Smith, A.B. Wilson, and Y. Yom-Tov. 2006. Ecotypic variation in the context of global climate change: Revisiting the rules. *Ecology Letters* 9:853–869.

Moore, F.R., R.J. Smith, and R. Sandberg. 2005. Stopover ecology of intercontinental migrants: En route problems and consequences for reproductive performance. In *Birds of Two Worlds*, ed. by R. Greenberg and P.P. Marra, 251–261. Washington, DC: Smithsonian Institution Press.

Murray, D.L., E.W. Cox, W.B. Ballard et al. 2006. Pathogens, nutrient deficiency, and climate influences on a declining moose population. *Wildlife Monographs* 166:1–30.

Myers, P., B.L. Lundrigan, S.M.G. Hoffman, A.P. Haraminac, and S.H. Seto. 2009. Climate-induced changes in the small mammal communities of the Northern Great Lakes Region. *Global Change Biology* 15:1434–1454.

Ostfeld, R.S. 1997. The ecology of Lyme-disease risk. *American Scientist* 85:338–46.

Parmesan, C. 2006. Ecological and evolutionary responses to recent climate change. *Annual Review of Ecology, Evolution, and Systematics* 37:637–669.

Parmesan, C., and G. Yohe. 2003. A globally coherent fingerprint of climate change impacts across natural systems. *Nature* 421:37–42.

Price, J., S. Droege, and A. Price. 1995. *The Summer Atlas of North American Birds.* London/San Diego: Academic Press.

Reále, D., A.G. McAdam, S. Boutin, and D. Berteaux. 2003. Genetic and plastic responses of a northern mammal to climate change. *Proceedings of the Royal Society of London Series B–Biological Sciences* 270:591–596.

Root, T.L. 1988. Environmental factors associated with avian distributional boundaries. *Journal of Biogeography* 15:489–505.

Root, T.L., J.T. Price, K.R. Hall, S.H. Schneider, C. Rosenzweig, and J.A. Pounds. 2003. Fingerprints of global warming on wild animals and plants. *Nature* 421:57–60.

Root, T.L., D.P. MacMynowski, M.D. Mastrandrea, and S.H. Schneider. 2005. Human modified temperatures induce species changes: Joint attribution. *Proceedings of the National Academy of Sciences* 102:7465–7469.

Rosenzweig, C., G. Casassa, DJ. Karoly et al. 2007. Assessment of observed changes and responses in natural and managed systems. In *Climate Change 2007: Impacts, Adaptation and Vulnerability*, ed. by M.L. Parry, O.F. Canziani, J.P. Palutikof, P.J. van der Linden, and C.E. Hanson, 79–131. Cambridge: Cambridge University Press.

Smith, R., M. Hamas, M. Dallman, and D. Ewert. 1998. Spatial variation in foraging of the Black-throated Green Warbler along the shoreline of northern Lake Huron. *Condor* 100:474–484.

Strode, P.K. 2003. Implications of climate change for North American wood warblers (*Parulidae*). *Global Change Biology* 9:1137–1144.

Thomas, C.D., E.J. Bodsworth, R.J. Wilson et al. 2001. Ecological and evolutionary processes at expanding range margins. *Nature* 411:577–581.

Thomas, C.D., A. Cameron, R.E. Green et al. 2004. Extinction risk from climate change. *Nature* 427:145–148.

Trenberth, K.E., P.D. Jones, P. Ambenje et al. 2007. Observations: Surface and atmospheric climate change. In *Climate Change 2007: The Physical Science Basis*, ed. by S. Solomon, D. Qin, M. Manning et al., 235–336. Cambridge and New York: Cambridge University Press.

Trumpickas, J., B.J. Shuter, and C.K. Minns. 2009. Forecasting impacts of climate change on Great Lakes surface water temperatures. *Journal of Great Lakes Research* 35:454–463.

Tucker, J.K., C.R. Dolan, J.T. Lamer, and E.A. Dustman. 2008. Climatic warming, sex ratios, and red-eared sliders (*Trachemys scripta elegans*) in Illinois. *Chelonian Conservation and Biology* 7:60–69.

Visser, M.E. 2008. Keeping up with a warming world: Assessing the rate of adaptation to climate change. *Proceedings of the Royal Society B–Biological Sciences* 275:649–659.

Visser, M.E., and C. Both. 2005. Shifts in phenology due to global climate change: The need for a yardstick. *Proceedings of the Royal Society B–Biological Sciences* 272:2561–2569.

Visser, M.E., L.J.M. Holleman, and P. Gienapp. 2006. Shifts in caterpillar biomass phenology due to climate change and its impact on the breeding biology of an insectivorous bird. *Oecologia* 147:164–172.

Vucetich, J.A., and R.O. Peterson. 2004. The influence of top-down, bottom-up, and abiotic factors on the moose (*Alces alces*) population of Isle Royale. *Proceedings of the Royal Society of London, Series B* 271:183–89.

Willis, C.G., B. Ruhfel, R.B. Primack, A.J. Miller-Rushing, and C.C. Davis. 2008. Phylogenetic patterns of species loss in Thoreau's woods are driven by climate change. *Proceedings of the National Academy of Sciences* 105:17029–17033.

Winder, M., and D.E. Schindler. 2004. Climate change uncouples trophic interactions in an aquatic ecosystem. *Ecology* 85:2100–2106.

Wilmers, C.C., E. Post, R.O. Peterson, and J.A. Vucetich. 2006. Predator disease out-break modulates top-down, bottom-up and climatic effects on herbivore population dynamics. *Ecology Letters* 9:383–389.

Discussion Questions

1   What are the five types of changes already observed among species in response to climate disruption?

2   What evidence in the Great Lakes region already documents the changing ranges of species?

3   What concerns are raised about species' phenology changing at different rates?

4   Figure 9.3 summarizes the array of climate factors upon species effects. Develop a scenario with representative species and ecosystems to illustrate some of the complexity of interactions.

5   Based on the eight factors in the "rules of thumb" for vulnerable species, investigate what species from the Great Lakes region are good candidates for adaptive management.

CHAPTER 10

Global-Scale Responses and Vulnerability of Marine Species and Fisheries to Climate Change[1]

William W. L. Cheung and Daniel Pauly

Of the various ways humans affect marine ecosystems, climate change may be the most insidious and unrecognized. In fact, even if they believe that it is occurring, most people think climate change is going to affect us later, and thus there is no real urgency. As we show below, however, climate change has already begun to affect us in multiple ways, including through effects on the oceans and marine fisheries. This chapter is thus devoted to documenting some of the work through which scattered observations on the effect of climate change on marine organisms were generalized and, in the process, the first global maps of observed and predicted climate change impacts on marine biodiversity and fisheries were produced, thus complementing work performed in the terrestrial realm.

Climate change affects ocean properties including water temperature, oxygen level, and acidity. According to the Intergovernmental Panel on Climate Change (IPCC) 5th Assessment Report (AR5), there is compelling evidence that the heat content and the stratification of the ocean have been increasing in the twentieth century, while sea-ice and pH have been decreasing, and that these trends can be expected to continue in the next century under the climate change scenarios considered by the IPCC (2014). Also, available evidence indicates that climate change is expected to result in expansion of oxygen minimum zones, changes in primary productivity, changes in ocean circulation patterns, sea level rises, and increase in extreme weather events.

Marine fishery catches consist almost solely of fishes and invertebrates that are biologically sensitive to changes in temperature, oxygen level, and other ocean conditions; thus, we expect that fisheries are being affected by climate change (CC) and ocean acidification (OA). In the ocean, physiological performance of aquatic and marine water-breathing organisms is strongly dependent on temperature and oxygen (Pauly 1981, 2010; Pörtner 2010). Because of the higher viscosity of water and because dissolved oxygen occurs in water in lower concentration than in the atmosphere, it is energetically costly for water-breathing organisms to obtain oxygen for respiration from water (Pauly 1981, 2010). Thus, changes in oxygen supply and demand are expected to have large implications for respiration and other body functions of fishes and invertebrates (i.e., water-breathing ectotherms). Although low oxygen tolerance thresholds vary across species and life stages, they tend to be highest for large water-breathing ectotherms. When temperature becomes either too high or too low, oxygen supply capacity decreases relative to oxygen demand and thus limits metabolism. The ranges between the lowest and highest temperatures that are tolerated by organisms are generally consistent with the variability of environmental temperatures they are generally exposed to and can change during their life cycles. Also, smaller individuals are more heat tolerant than large ones, in line with observations of declining animal body sizes in warming oceans (Daufresne et al. 2009). Increases in CO_2 in the ocean may also have direct and indirect effects on growth, reproduction, and survivorship of fishes and invertebrates, particularly those that form calcium carbonate exoskeleton (Kroeker et al. 2013).

An understanding of the physiological sensitivity and responses to ocean temperature, oxygen, acidity, and other water properties allows us to develop hypotheses about how climate change and ocean acidification will affect exploited fish stocks and fisheries. Theses hypotheses include the following:

- Given ocean warming, fishes and invertebrates will be shifting their distributions, mainly to higher latitude and deeper water to maintain their thermal niche.

- In nontropical systems, warmer-water species will increase their contribution to local catches.

- Maximum body size of fishes will decrease as the oceans become warmer and less oxygenated.

- Global marine catches will decline, particularly in the tropics.

To examine these hypotheses at the global scale, we conducted a series of empirical and theoretical studies that made use of species' distribution ranges (see chapter 4) and spatially explicit global catch data from the *Sea Around Us* (see chapter 5) to evaluate the effects of climate change and ocean acidification on the distribution of exploited species, the species composition of catches, and projected fisheries catch potential.

As a conceptually first step (though one taken last in the sequence of studies described below), the signature of the effects of ocean temperature change on species composition of fisheries catches was studied, using a newly developed metric called mean temperature of the catch (MTC). The second step was to develop a species distribution model, called the dynamic bioclimate envelope model (DBEM), that predicts changes in the distribution ranges of exploited marine species and the patterns of species richness in response to changing ocean conditions. Once developed, the DBEM was modified so as to progressively account for an increasing number of features, such as the population dynamics of the species included, their dispersal modes, interactions with other species, association with different habitats, oxygen requirements, and resistance to low pH. The third step was to use macroecological theory to derive the theoretical relationship between net primary production (NPP), biogeography, and fisheries catch potential and to express that relationship in a single empirical equation. Then, with projected changes in NPP and species distributions combined, future changes in distribution of fisheries catch potential and the maximum body size of exploited species could be projected. Finally, by use of the DBEM and basic principles of geometry and physiology, the effects of ocean warming and deoxygenation on the maximum body size of exploited fishes could be projected.

Mean Temperature of the Catch

Marine fishes and invertebrates exhibit physiological thermal tolerances that constrain them to live within a certain range of temperatures. Thus, for example, the seasonal migration of fishes up and down along the coast of northwest Africa tracks the seasonal temperature oscillations along that same coast (Figure 10.1). Similarly, as the oceans warm up, fishes and invertebrates have to shift their distributions in order to maintain themselves in habitats with their preferred temperature. This results (at locations outside of the tropics) in changes in species composition, as the taxa increase in abundance that are adapted to warmer waters. For example, warming in the eastern Mediterranean (i.e., in Greece) caused a reduction of bearded horse mussel *Modiulus barbatus* (Katsikatsou et al. 2011) and the establishment of the noxious silver-cheeked toadfish Lagocephalus sceleratus, a Red Sea or "Lessepsian" migrant (Kasapidis et al. 2007). Shifted species distribution ranges follow temperature clines from high to low, reflecting a lateral gradient at the basin scale (Pinsky et al. 2013; Poloczanska et al. 2013) or a vertical temperature gradient to deeper waters (Dulvy et al. 2008; Pauly 2010). However, the implications of such responses for global fisheries to ocean warming had not been empirically demonstrated.

The newly developed index, the MTC, shows that global catches are increasingly dominated by warmer-water species (Cheung et al. 2013c). The MTC is the weighted average of the preferred temperatures of the various fish and invertebrate species in the catch. The preferred temperature of each species (which is expected to be fairly stable in evolutionary time for most taxa) was predicted from overlaying the current distribution of the species, as mapped using the approach described in chapter 4 and sea surface temperature (SST). Therein, species that are distributed in warmer waters will have higher preferred temperature and vice versa. Thus, for example, if the catch of a small country in the temperate zone is increasingly dominated by warmer-water species, its MTC would increase.

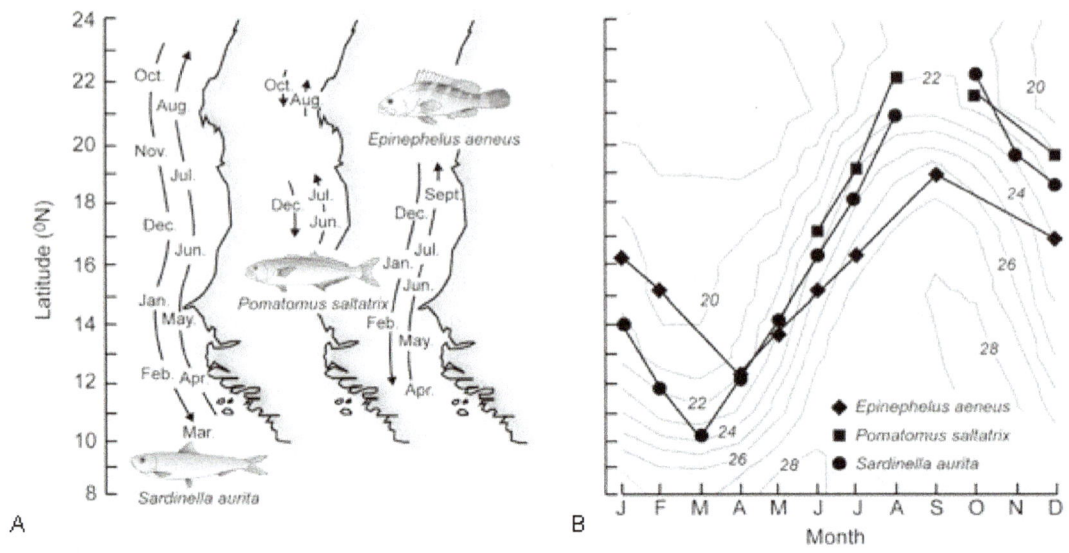

Figure 10.1 Seasonal latitudinal migrations of some northwest African fishes. (A) Summary of information on the occurrence in space (latitude) and time (month) of 3 species, *Sardinella aurita*, *Pomatomus saltator*, and *Epinephelus aeneus* (from Boëly et al. 1978, 1979; Champagnat and Domain 1978; Barry-Gérard 1994). (B) Same as in (A) but plotted against mean monthly temperature. Data from the Comprehensive Ocean–Atmosphere Data Set (COADS). The seasonal migrations result in the 3 species remaining in approximately the same temperature range (and hence having the same oxygen consumption) throughout the year. (Adapted from Pauly 1994.)

Using the *Sea Around Us* catch data, the MTC was calculated for all the Large Marine Ecosystems (LMEs) of the world from 1970 to 2006. After the effects of fishing and large-scale oceanographic variability were accounted for, global MTC increased at a rate of 0.19°C per decade between 1970 and 2006, and nontropical MTC increased at a rate of 0.23°C per decade. In tropical areas, MTC increased initially because of the reduction in the proportion of subtropical species catches but subsequently stabilized as scope for further "tropicalization" of communities became limited (Figure 10.2). By showing that changes in MTC are significantly related to changes in SST across LMEs, Cheung et al. (2013c) showed conclusively that ocean warming has already affected global fisheries catch composition in the past four decades. This is now being verified at smaller scales (Keskin and Pauly 2014; Tsikliras and Stergiou 2014).

Projecting Distribution Shifts of Exploited Species

Given that changes in the composition of fisheries catches are probably being driven by warming-induced biogeographic shifts, the next step was to investigate whether exploited species would continue to shift their biogeography in the future under climate change and how

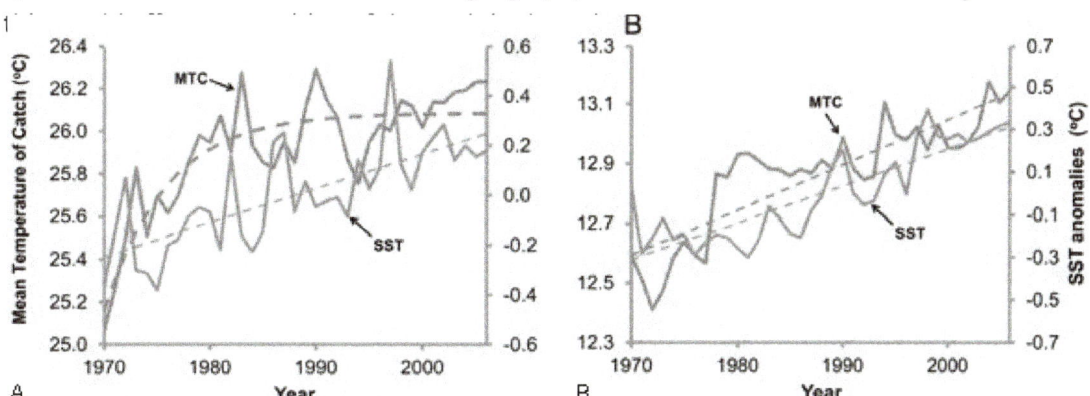

Figure 10.2   Observed trends in mean temperature of the catch (MTC) and sea surface temperature (SST) from (A) tropical and (B) nontropical Large Marine Ecosystems. (Adapted from Cheung et al. 2013c.)

A DBEM was developed to project future distribution of more than 1,000 exploited fishes and invertebrates globally. The DBEM, described in Cheung et al. (2008b, 2009) and later in Cheung et al. (2011) and Fernandes et al. (2013), predicts a species' range (on a grid of about 180,000 cells of 30′ latitude by 30′ longitude representing the world ocean) based on the association between the modeled distributions and environmental variables. The current distribution of a taxon is predicted using a method developed by the *Sea Around Us* and described in chapter 4. Comparison with other species distribution modeling approaches that allow prediction of distributions for all the exploited species show that the DBEM performs equally well in terms of test statistics with observed occurrence data, for example, receiver operating characteristics (Jones et al. 2012). The DBEM infers preference profiles, defined as the suitability of different environmental conditions to the species covered, from their predicted current distributions, assuming that species' current distributions match their environmental preference. Distinguishing features of the DBEM relative to other species distribution models include the explicit representation of spatial population dynamics (Cheung et al. 2008b), ecophysiology (Cheung et al. 2011) and, in a new version of DBEM, trophic interactions (Fernandes et al. 2013).

We applied the DBEM to project future distributions of 1,066 species of exploited fishes and invertebrates under climate change scenarios developed by the IPCC. These species include the overwhelming majority of the taxa whose population is large enough to generate catches that are reported at the species level in the global fisheries statistics of the Food and Agriculture Organization of the United Nations (FAO) and thus represent a very large sample of marine macrofauna. The rate of range shift and the intensity of species invasion and local extinction in the global ocean by 2050 relative to the 2000s were then calculated.

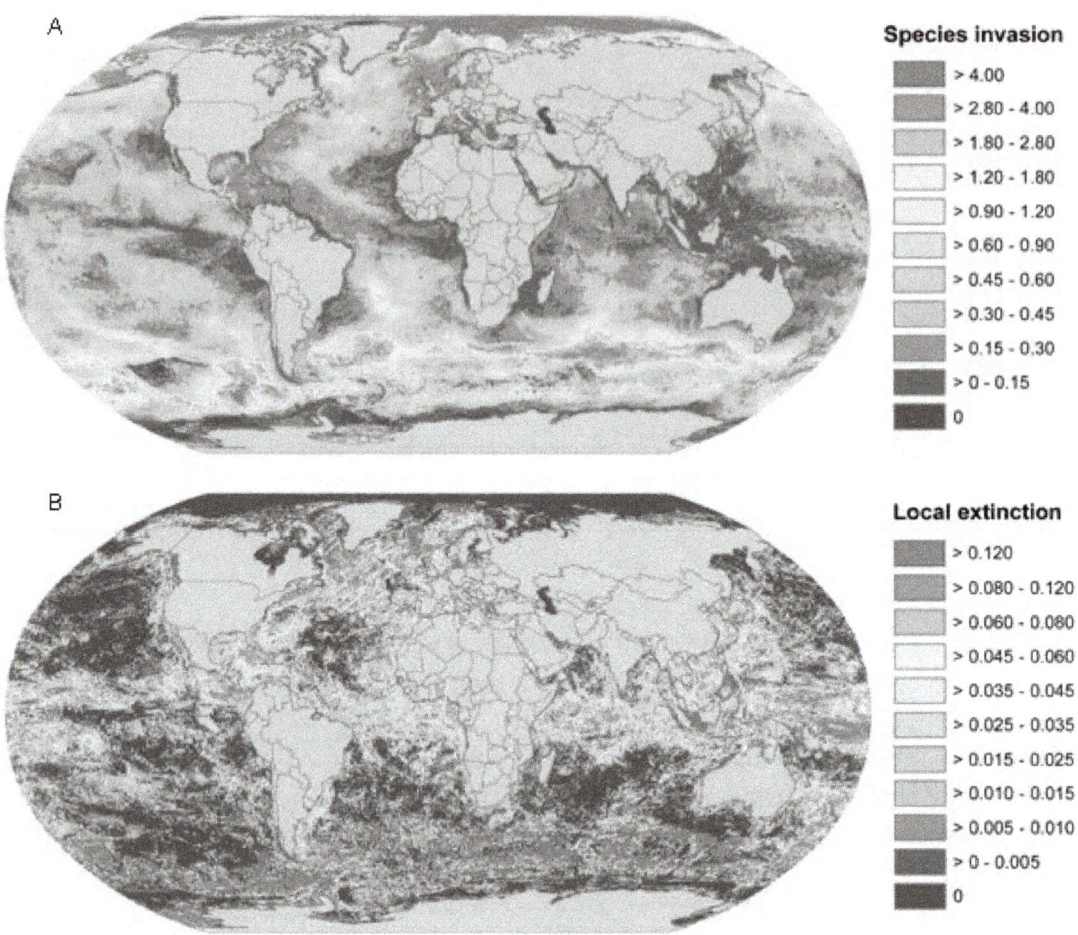

Figure 10.3 **Projected intensity of (A) species invasion and (B) local extinctions by 2050 relative to 2000 (10-year average) under the IPCC Special Report on Emission Scenarios (SRES) A1B scenario. (Adapted from Cheung et al. 2009.)**

The resulting projections show that climate change leads, overall, to range shifts to higher latitude and deeper waters (Figure 10.3), although some species display range shift in opposite directions as they follow local rather than large-scale climate change gradients (Cheung et al. 2009; Pinsky et al. 2013). Thus, numerous local extinctions in the subpolar regions, the tropics, and semienclosed seas can be expected. Simultaneously, species invasions are projected to be most frequent in the Arctic and the Southern Ocean. Jointly, these results, which are robust to the selection of species distribution models (Jones and Cheung 2015), suggest a dramatic turnover of current marine biodiversity, implying ecological disturbances that will massively disrupt the provision of ecosystem services. Moreover, these results support the hypothesis that the observed pattern of changes in species composition of catches, as indicated by the MTC introduced above, will continue in the future.

Relationship between NPP and Maximum Catch Potential

We developed an empirical relationship to predict maximum catch potential of a species based on NPP and the biogeography and ecology of marine water-breathers.

First, based on theories linking trophic energetics and allometric scaling of metabolism, a theoretical model was developed that relates the maximum catch potential of a species to its trophic level, geographic range, and mean NPP within the species' exploited range (Cheung et al. 2008a). Therein, the relationship between metabolic rate and body size of marine organisms, as quantified by an allometric equation and the energy available for a specific population of animals at trophic level (λ) in an ecosystem, was calculated based on a trophic transfer efficiency (TE) set at 10% (Pauly and Christensen 1995; Ware 2000). Finally, maximum sustainable yield (MSY) of the population was obtained from $rB_\infty/4$, where B_∞ is the biomass at carrying capacity and r is the intrinsic rate of population increase, as implied in Schaefer (1954) and stated explicitly in Ricker (1975, p. 315) and Pauly (1984a, p. 140). This led to a theoretical relationship in which MSY was expressed as a function of NPP, λ, and TE.

This model was fitted to catch, ecological, and biogeographical data from 1,000 species of exploited marine fish and invertebrate species. This led to an empirical relationship (i.e., a multiple regression) between these species' approximated maximum catch potential and their ecology and biogeography. Therein, maximum catch potentials were assumed to be approximated by the average of their five highest annual catches (from 1950 and 2006), nearly the same assumption as made by Srinivasan et al. (2010). Additional variables were included in the empirical model to correct for biases resulting from the uncertainty inherent in the original catch data. The empirical model had a high explanatory power ($R^2 = 0.703$), and the signs and magnitudes of the partial slopes agreed with theoretical expectations. Friedland et al. (2012) suggest that chlorophyll A concentration is a better predictor of catch potential, but reanalysis of Cheung et al. (2008a) using chlorophyll A concentration instead of NPP does not result in a significantly better model. Thus, the empirical model can be combined with the DBEM to project the impacts of climate change on global marine fisheries, our next topic.

Projecting Future Catch Potential

By applying the empirical model described in the last section to projected future distributions derived from the DBEM, we could then estimate changes in global maximum catch potential for 1,066 species of exploited marine fish and invertebrates from 2005 to 2055 (Cheung et al.

Figure 10.4 Projected change in maximum catch potential by 2050 relative to 2005 (10-year average) under the SRES A1B scenario. (Adapted from Cheung et al. 2010.)

2010). Species distribution projections were obtained from Sarmiento et al. (2004), working with the IPCC Special Report on Emission Scenarios (SRES) A1B and B2, along with projected changes in NPP. Once this information was incorporated into the empirical equation, it appeared that climate change would lead to a large-scale redistribution of global catch potential, with an average of 30%–70% increase in high-latitude regions and a drop of up to 40% in the tropics (Figure 10.4). Moreover, predicted maximum catch potential declined in the lower-latitude margins of semienclosed seas, while it increased near the poleward tips of continental shelf margins, particularly in the Pacific Ocean. Among the twenty Exclusive Economic Zones (EEZs) with the highest landings according to the FAO, those with the highest increase in catch potential by 2055 belonged to Norway, Greenland, the United States (Alaska), and the Russian Far East. In contrast, the catch potential of Indonesia, the contiguous United States, Chile, and China have EEZs whose catch potential was predicted to decline most strongly. These results highlight the need to develop adaptation and mitigations policies for climate change impacts on fisheries, particularly in the tropics.

Because the above studies did not account for the effects of changes in ocean biogeochemistry and phytoplankton community structure, a version of the DBEM was developed that incorporated these factors and was used to project the distributions and maximum catch potentials of 120 species of exploited demersal fish and invertebrates in the northeast Atlantic (Cheung et al. 2011). Using projections from the U.S. National Oceanic and Atmospheric Administration's Geophysical Fluid Dynamics Laboratory Earth System Model (ESM2.1) under the SRES A1B, we predicted ocean acidification and reduction in oxygen content to reduce growth performance, increase the rate of range shift, and lower the estimated catch potentials (10-year average of 2050 relative to 2005) by 20%–30% relative to simulations that did not consider these factors (Figure 10.5). Consideration of changes in phytoplankton community structure may further reduce projected catch potentials by 10%. These results highlight the sensitivity of marine ecosystems to biogeochemical changes and the need to incorporate likely hypotheses of their biological and ecological effects in assessing climate change impact

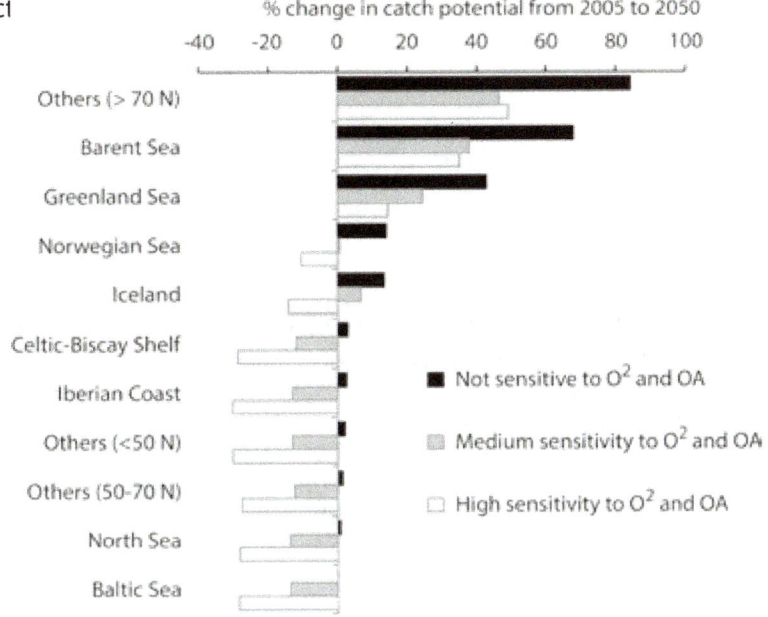

Figure 10.5   Projected change in maximum catch potential in Large Marine Ecosystems in the northeast Atlantic by 2050 relative to 2005 (10-year average) under the SRES A2 scenario. OA, ocean acidification. (Adapted from Cheung et al. 2011.)

Projecting Maximum Body Size of Fish and Invertebrates

Both theory and empirical observations support the hypothesis that warming and reduced oxygen will reduce body size of marine fishes (Pauly 1998a) and invertebrates (Pauly 1998b). Changes in temperature, oxygen content, and other ocean biogeochemical properties directly affect the ecophysiology of marine water-breathing organisms. Particularly, their physiological performance, including their growth rate and size at first reproduction, are strongly dependent on temperature and oxygen (Pauly 1981, 1984a, 1984b, 1998b, 2010; Pörtner and Farrell 2008). An organism's low oxygen tolerance threshold varies across species, body size, and life stage and is highest for large organisms. The oxygen tolerance threshold is set by the capacity of an organism's respiratory and circulatory systems to supply O_2 and cover demand. A corollary of the above is that distribution, growth, size at first reproduction, maximum body size, and survival of fishes are controlled by the balance between oxygen supply and demand under different temperatures (Pütter 1920; Pauly 1981, 2010; Kolding et al. 2008). As fishes increase in size

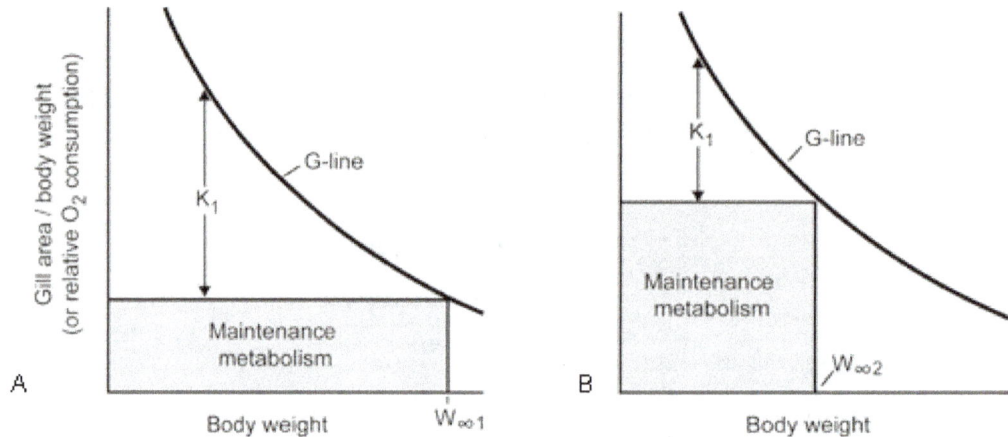

Figure 10.6 Diagram illustrating how maintenance metabolism determines asymptotic weight (W_∞), given a "G-line" defined by the growth of the gills relative to body weight, because at W_∞, relative gill area (and hence oxygen supply) is just enough for maintenance metabolism (shaded areas). (A) Fish exposed to a low level of stress (e.g., low temperature, abundant oxygen, abundant food). (B) Fish exposed to a higher level of stress (low oxygen concentration, high temperature, causing rapid denaturation of body protein, or low food density, requiring O_2 to be diverted to foraging rather than protein synthesis). Note that food conversion efficiency and hence also the scope for growth are directly related to the distance, in these graphs, between the G-line and the level of maintenance metabolism (see Pauly, 1981, 1984b, 2010). (Adapted from Pauly 2010.)

(weight), mass-specific oxygen demand increases more rapidly than oxygen supply (Pauly 1997). Thus, fish reach maximum body size when oxygen supply is balanced by oxygen demand (Figure 10.6A). Moreover, the scope for aerobic respiration and growth decreases when size increases, that is, oxygen supply per unit body weight decreases (Figure 10.6B). The decrease in food conversion efficiency that this implies decreases the biomass production of fish and invertebrate populations.

However, although the interrelationships of temperature, oxygen, and growth in water-breathers are well established in the laboratory (Pauly 2010), the extent to which maximum body size of fishes would be affected by projected changes in temperature and oxygen level in the oceans remained unexplored. The DBEM was thus used to examine the integrated biological responses of more than 600 species of marine fishes to changes in distribution, abundance, and body size (Cheung et al. 2013a, 2013b), based on explicit representations of ecophysiology, dispersal, distribution, and population dynamics. The result was that assemblage-averaged maximum body weight is expected to decline by 14%–24% globally from 2000 to 2050 under a high-emission scenario (Figure 10.7). The projected magnitude of decrease in body size is consistent with experimental (Forster et al. 2012; Cheung et al. 2013b) and field observations (Baudron et al. 2014). About half of this shrinkage is caused by changes in distribution and abundance, the remainder by changes in physiological performance. The tropical and intermediate latitudinal areas will be heavily affected, with an average reduction of more than 20%. Decreases in growth and body size should reduce the biomass production of fish populations, and hence fishery catches, and potentially alter trophic interactions.

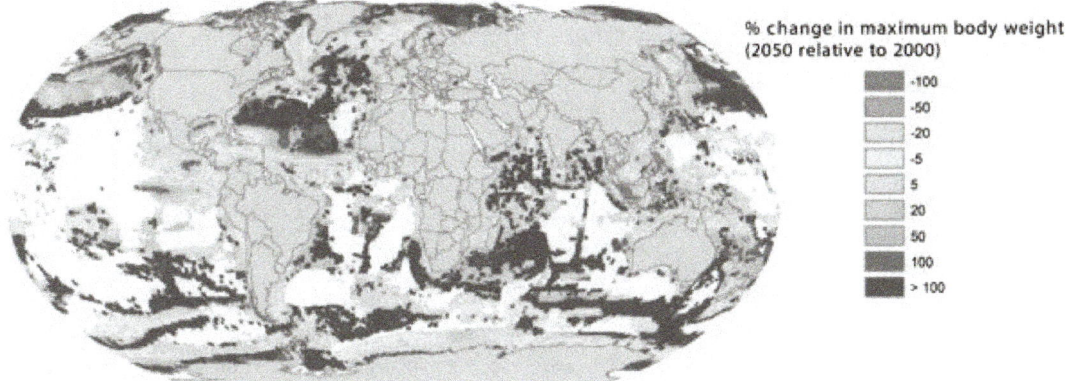

Figure 10.7 Projected change in maximum body weight of exploited fishes by 2050 relative to the 2000 period (20-year average) under the SRES A2 scenario. (Adapted from Cheung et al. 2013b.)

Conclusion

In their series of studies, the authors and their collaborators detected a signature of ocean warming on global fisheries in the last four decades, and they also projected that such changes would continue over the next 40 years, leading to strong species turnover, redistribution of fisheries catch potential, and decreases in the maximum body sizes of exploited species of fishes and invertebrates. Results from these global-scale analyses highlighted the inequity of climate change impacts to different regions of the world. Specifically, the tropics will be affected by a high rate of local species extinction, a decrease in catch potential, and a larger decrease in body size of fishes. Many tropical communities are highly dependent on local fisheries for food and livelihood (e.g., Zeller et al. 2015), but their economic and societal capacity to adapt to climate change impacts on fisheries is often low. Thus, tropical fisheries are highly vulnerable to climate change, although tropical countries contribute little to the greenhouse gas emissions that cause climate change.

Future studies should address additional challenges to detecting, attributing, and projecting climate change and ocean acidification impacts on marine fisheries. First, the adaptive scope of exploited marine species and their fishers to impacts from climate change and ocean acidification should be evaluated. Second, different modeling approaches in projecting future seafood production under climate change and ocean acidification should be tested, to assess the utility of these approaches and quantify the level of uncertainty associated with the model projections.

Third, the effects of multiple stressors (i.e., climate and nonclimate) and their interactions must be explored. In addition, more regional studies to downscale the global analyses must be conducted, through which the weaknesses associated with the coarse projections of ocean properties from global circulation models can be better addressed. Such regional-scale analyses are more useful for informing national fisheries and coastal management agencies, which will both be challenged by global warming in coming years.

[...]

References

Barry-Gérard, M. 1994. Migration des poissons le long du littoral sénégalais. Pp. 215–234 in M. Barry-Gérard, T. Diouf, and A. Fonteneau (eds.), *L'évaluation des ressources exploitables par la pêche sénégalaise.* Édition ORSTOM, Paris.

Baudron, A. R., C. L. Needle, A. D. Rijnsdorp, and C. Tara Marshall. 2014. Warming temperatures and smaller body sizes: synchronous changes in growth of North Sea fishes. *Global Change Biology* 20: 1023–1031.

Boëly, T. 1979. Biologie de deux espèces de Sardinelles (*Sardinella aurita* Valenciennes 1887 et *Sardinella maderensis* Lowe 1841) des côtes sénégalaise. *Pêches Maritimes* July 1979: 426–430.

Boëly, T., J. Chabanne, and P. Fréon. 1978. Schéma migratoire de poissons pélagiques côtiers dans la zone sénégalo-mauritanienne. Pp. 63–70 in *Rapport du groupe de travail ad hoc sur les poissons pélagiques côtiers ouest africains de la Mauritanie au Libéria (26 °N à 5 °N).* FAO/Comité des pêches pour l'Atlantique. Centre-Est: COPACE/RACE Sér. 78/10.

Champagnat, C., and F. Domain. 1978. Migrations des poissons démersaux le long des côtes ouest africaines de 10 °N à 24 °N. *Cahier ORSTOM, série Océanographie* 16(3–4): 239–261.

Cheung, W. W. L., C. Close, V. W. Y. Lam, R. Watson, and D. Pauly. 2008a. Application of macroecological theory to predict effects of climate change on global fisheries potential. *Marine Ecology Progress Series* 365: 187–197.

Cheung, W. W. L., J. Dunne, J. L. Sarmiento, and D. Pauly. 2011. Integrating ecophysiology and plankton dynamics into projected maximum fisheries catch potential under climate change in the northeast Atlantic. *ICES Journal of Marine Science* 68: 1008–1018.

Cheung, W. W. L., V. W. Y. Lam, and D. Pauly. 2008b. Dynamic bioclimate envelope model to predict climate-induced changes in distributions of marine fishes and invertebrates. Pp. 5–50 in W. W. L. Cheung, V. W. Y. Lam, and D. Pauly (eds.), *Modelling present and climate-shifted distribution of marine fishes and invertebrates.* Fisheries Centre Research Reports 16(3), University of British Columbia, Vancouver, Canada.

Cheung, W. W. L., V. W. Y. Lam, J. L. Sarmiento, K. Kearney, R. Watson, and D. Pauly. 2009. Projecting global marine biodiversity impacts under climate change scenarios. *Fish and Fisheries* 10: 235–251.

Cheung, W. W. L., V. W. Y. Lam, J. L. Sarmiento, K. Kearney, R. E. G. Watson, D. Zeller, and D. Pauly. 2010. Large-scale redistribution of maximum fisheries catch potential in the global ocean under climate change. *Global Change Biology* 16: 24–35.

Cheung, W. W. L., D. Pauly, and J. L. Sarmiento. 2013a. How to make progress in projecting climate change impacts. *ICES Journal of Marine Science* 70: 1069–1074.

Cheung, W. W. L., J. L. Sarmiento, J. Dunne, T. L. Frölicher, V. Lam, M. L. D. Palomares, R. Watson, and D. Pauly. 2013b. Shrinking of fishes exacerbates impacts of global ocean changes on marine ecosystems. *Nature Climate Change* 3: 254–258.

Cheung, W. W. L., R. Watson, and D. Pauly. 2013c. Signature of ocean warming in global fisheries catches. *Nature* 497: 365–368.

Daufresne, M., K. Lengfellner, and U. Sommer. 2009. Global warming benefits the small in aquatic ecosystems. *Proceedings of the National Academy of Sciences* 106: 12788–12793.

Dulvy, N. K., S. I. Rogers, S. Jennings, V. Stelzen-müller, S. R. Dye, and H. R. Skjoldal. 2008. Climate change and deepening of the North Sea fish assemblage: a biotic indicator of warming seas. *Journal of Applied Ecology* 45: 1029–1039.

Fernandes, J. A., W. W. Cheung, S. Jennings, M. Butenschön, L. Mora, T. L. Frölicher, M. Barange, and A. Grant. 2013. Modelling the effects of climate change on the distribution and production of marine fishes: accounting for trophic interactions in a dynamic bioclimate envelope model. *Global Change Biology* 19(8): 2596–2607.

Forster, J., A. G. Hirst, and D. Atkinson. 2012. Warming-induced reductions in body size are greater in aquatic than terrestrial species. *Proceedings of the National Academy of Sciences* 109: 19310–19314.

Friedland, K. D., C. Stock, K. F. Drinkwater, J. S. Link, R. T. Leaf, B. V. Shank, J. M. Rose, C. H. Pilskaln, and M. J. Fogarty. 2012. Pathways between primary production and fisheries yields of large marine ecosystems. *PLoS ONE* 7: e28945.

IPCC. 2014. Summary for policymakers. In *Climate Change 2014: Impacts, Adaptation, and Vulnerability. Part A: Global and Sectoral Aspects*. Contribution of Working Group II to the Fifth Assessment Report of the Intergovernmental Panel on Climate Change [Field, C. B., V. R. Barros, D. J. Dokken, K. J. Mach, M. D. Mastrandrea, T. E. Bilir, M. Chatterjee, K. L. Ebi, Y. O. Estrada, R. C. Genova, B. Girma, E. S. Kissel, A. N. Levy, S. MacCracken, P. R. Mastrandrea, and L. L. White (eds.)]. Cambridge University Press, Cambridge, England.

Jones, M. C., and W. W. L. Cheung. 2015. Multi-model ensemble projections of climate change effects on global marine biodiversity. *ICES Journal of Marine Science* 72(3): 741–752.

Jones, M., S. Dye, J. Pinnegar, R. Warren, and W. W. L. Cheung. 2012. Modelling commercial fish distributions: prediction and assessment using different approaches. *Ecological Modelling* 225: 133–145.

Kasapidis, P., P. Peristeraki, G. Tserpes, and A. Magoulas. 2007. First record of the Lessepsian migrant *Lagocephalus sceleratus* (Gmelin 1789) (Osteichthyes: Tetraodontidae) in the Cretan Sea (Aegean, Greece). *Aquatic Invasions* 2(1): 71–73.

Katsikatsou, M., A. Anestis, H. O. Pörtner, T. Kampouris, and B. Michaelidis. 2011. Field studies on the relation between the accumulation of heavy metals and metabolic and HSR in the bearded horse mussel *Modiolus barbatus*. *Comparative Biochemistry and Physiology Part C: Toxicology & Pharmacology* 153: 133–140.

Keskin, Ç., and D. Pauly. 2014. Changes in the "mean temperature of the catch": application of a new concept to the north-eastern Aegean Sea. *Acta Adriatica* 55(2): 213–218.

Kolding, J., L. Haug, and S. Stefansson. 2008. Effect of ambient oxygen on growth and reproduction in Nile tilapia (*Oreochromis niloticus*). *Canadian Journal of Fisheries and Aquatic Science* 65: 1413–1424.

Kroeker, K. J., R. L. Kordas, R. Crim, I. E. Hendriks, L. Ramajo, G. S. Singh, C. M. Duarte, and J. P. Gattuso. 2013. Impacts of ocean acidification on marine organisms: quantifying sensitivities and interaction with warming. *Global Change Biology* 19: 1884–1896.

Pauly, D. 1981. The relationship between gill surface area and growth performance in fish: a generalization of von Bertalanffy's theory of growth. *Berichte der Deutschen Wissenschaftlichen Kommission für Meeresforschung* 28: 251–282.

Pauly, D. 1984a. *Fish population dynamics in tropical waters: a manual for use with programmable calculators*. ICLARM Studies and Reviews 8.

Pauly, D. 1984b. A mechanism for the juvenile-to-adult transition in fishes. *Journal du Conseil International pour l'Exploration de la Mer* 41: 280–284.

Pauly, D. 1994. Un mécanisme explicatif des migrations des poissons le long des côtes du Nord-Ouest africain. Pp. 235–244 in M. Barry-Gérard, T. Diouf, and A. Fonteneau (eds.), *L'évaluation des ressources exploitables par la pêche artisanale sénégalaise*. Documents scientifiques présentés lors du Symposium, 8–13 février 1993, Dakar, Sénégal. ORSTOM Éditions, Paris.

Pauly, D. 1997. Geometrical constraints on body size. *Trends in Ecology and Evolution* 12: 442–443.

Pauly, D. 1998a. Tropical fishes: patterns and propensities. Pp. 1–17 in T. E. Langford, J. Langford, and J. E. Thorpe (eds.), Tropical fish biology. *Journal of Fish Biology* 53(Suppl. A).

Pauly, D. 1998b. Why squids, though not fish, may be better understood by pretending they are. Pp. 47–58 in A. I. L. Payne, M. R. Lipinski, M. R. Clarke, and M. A. C. Roeleveld (eds.), Cephalopod biodiversity, ecology and evolution. *South African Journal of Marine Science* 20.

Pauly, D. 2010. *Gasping Fish and Panting Squids: Oxygen, Temperature and the Growth of Water-Breathing Animals*. Excellence in Ecology Series Vol. 22, International Ecology Institute, Oldendorf/Luhe, Germany.

Pauly, D., and V. Christensen. 1995. Primary production required to sustain global fisheries. *Nature* 374: 255–257.

Pinsky, M. L., B. Worm, M. J. Fogarty, J. L. Sarmiento, and S. A. Levin. 2013. Marine taxa track local climate velocities. *Science* 341: 1239–1242.

Poloczanska, E. S., C. J. Brown, W. J. Sydeman, W. Kiessling, D. S. Schoeman, P. J. Moore, K. Brander, J. F. Bruno, L. B. Buckley, M. T. Burrows, C. M. Duarte, B. S. Halpern, J. Holding, C. V. Kappel, M. I. O'Connor, J. M. Pandolfi, C. Parmesan, F. Schwing, S. A. Thompson, and A. J. Richardson. 2013. Global imprint of climate change on marine life. *Nature Climate Change* 3: 919–925.

Pörtner, H. O. 2010. Oxygen- and capacity-limitation of thermal tolerance: a matrix for integrating climate-related stressor effects in marine ecosystems. *Journal of Experimental Biology* 213: 881–893.

Pörtner, H. O., and A. P. Farrell. 2008. Physiology and climate change. *Science* 322: 690–692.

Pütter, A. 1920. Studien über physiologische Ähnlichkeit. VI. Wachstumsähnlichkeiten. *Pflüger's Archiv für die gesamte Physiologie* 180: 298–340.

Ricker, W. E. 1975. Computation and interpretation of biological statistics of fish populations. *Bulletin of the Fisheries Research Board of Canada* 191.

Sarmiento, J. L., R. Slater, R. Baber, L. Bopp, S. C. Doney, A. C. Hirst, J. Kleypas, R. Matear, U. Mikolajewicz, P. Monfray, V. Soldatov, S. A. Spall, and R. Stouffer. 2004. Response of ocean ecosystems to climate warming. *Global Biogeochemical Cycles* 18(3). doi: 10.1029/2003GB002134.

Schaefer, M. B. 1954. Some aspects of the dynamics of populations important to the management of the commercial marine fisheries. *Bulletin of the InterAmerican Tropical Tuna Commission* 1: 27–56.

Srinivasan, U. T., W. W. L. Cheung, R. Watson, and U. R. Sumaila. 2010. Food security implications of global marine catch losses due to overfishing. *Journal of Bioeconomics* 12(3): 183–200.

Tsikliras, A. C., and K. I. Stergiou. 2014. Mean temperature of the catch increases quickly in the Mediterranean Sea. *Marine Ecology Progress Series* 515(2014): 281–284.

Ware, D. M. 2000. Aquatic ecosystems: properties and models. Pp. 161–194 in P. J. Harrison and T. R. Parsons (eds.), *Fisheries Oceanography: An Integrative Approach to Fisheries Ecology and Management*. Blackwell Science, Oxford.

Zeller, D., S. Harper, K. Zylich, and D. Pauly. 2015. Synthesis of under-reported small-scale fisheries catch in Pacific-island waters. *Coral Reefs* 34(1): 25–39.

Discussion Questions

1   Since fish have preferred temperatures, what do they do as the seasons cause changes in water temperature? Describe an example from the chapter.

2   What part of the ocean is expected to have a decline in total fish catch?

3   The mean temperature of the catch (MTC) provides a measure of the preferred water temperature for the species caught by fishermen in a location. Why is it useful to know this in predicting the fate of marine fisheries?

4   Summarize the geographic changes in fish distribution and what is expected to happen to the fish themselves (e.g., abundance and size). What are the implications for societies that depend upon marine fish?

PART IV

AN ECONOMIST'S
PERSPECTIVE—REALITY
AND HOPE

CHAPTER 11

How Ideas Change Over Time

Nicholas Stern

Over the last three decades the world has experienced extraordinary changes—in income and growth, in health and life expectancy, in urbanization and demography, in technology and commerce. We have seen radical change in the balance of economic activity toward emerging-market and developing countries. And we are in the midst of an information and communication technology revolution that is upending old practices and modes of social interaction. These changes have brought immense benefits, together with, in some cases, social and economic tension and stress. They have also placed enormous pressures on natural resources and the environment, including the air we breathe, the water we drink, and the land we use. Having been fueled by hydrocarbons, these changes are also disrupting the climatic conditions within which our civilizations have developed.

Many of the profound changes of the last three decades will continue over the next three. We will see a fundamental structural transformation of the world economy. Global population is expected to reach around 9 billion by 2050, and more than 6 billion of those people will likely live in cities, which are the engines of global economic growth. Pressures on natural resources and the climate will intensify, increasingly threatening the potential for growth and prosperity, if we continue with old technologies and ways of doing things. At the same time, newer technologies are opening up extraordinary possibilities across the whole economy, including in the way we can generate, manage, and use energy.

Most of the decisions we need to take on climate change will be made in the context of these unfolding changes. Developing and middle-income countries are increasingly recognizing the unattractiveness of dirty models of industrial growth and the widespread benefits of a cleaner, more efficient, smarter-technology and service-oriented alternative. Restructuring toward cleaner and more efficient growth also provides a compelling response to many of the challenges faced by high-income countries, including weakening competitiveness, falling living standards of many, aging population, aging infrastructure, congested cities, and rising budget deficits. If the changes involved in the structural transformations are managed well, radically reducing waste, congestion, pollution, and the degrading or destruction of land and forests, the majority of the emissions reductions needed to stay within a 2°C pathway (with a 50–50 chance) could be delivered. Achieving the further reductions in emissions that are necessary will involve more ambitious policies and investments in the areas of energy systems, land use, and urban infrastructure, but if done wisely these too will bring many attractive economic, social, and environmental benefits.

The deepening understanding of these interlinked transformations that has emerged in recent years, and of the scale of the opportunities they bring, is one of the most important developments that has occurred since I led *The Stern Review* in 2005–2006. Indeed, the world has changed dramatically not just over the last 30 years but in these last few. There have been other important changes, too—mostly helpful, but some not so. We have experienced a major global financial crisis and recession, which have diverted many leaders' attention from climate change. Yet the costs of low-carbon technologies, particularly renewable energy technologies, have fallen dramatically over this period, making low-carbon substitutes in many parts of the world competitive with high-carbon incumbents. In particular, low-cost, decentralized renewable energy and storage technologies provide life-changing opportunities for many of the world's poor people, especially those without access to grid-supplied electricity—"sustainable energy for all" is becoming a realistic vision, not just an inspiring one. Yet technology has also advanced in hydrocarbons, making ever deeper and farther-flung fossil fuel deposits more accessible at a time when their use needs to be phased down.

The world has also gained much experience in climate policymaking at all levels of government, yielding valuable lessons for tomorrow's policymakers. At the international level, cooperation on climate change is moving—slowly, but moving—toward a more dynamic and collaborative model. At the domestic level, there is increasing recognition of the multifaceted role that policies and institutions can play in enabling a structural transformation toward low-carbon while promoting growth and poverty reduction. Cities are showing great leadership in reducing emissions through innovative approaches. And the important interconnections between climate change and other issues are becoming much clearer—not merely the linkages between climate impacts and other challenges, but also between policies to tackle climate change and the measures needed to tackle other significant challenges that countries face today. The climate-and-health linkages, for example, have been much more deeply studied and communicated to policymakers in recent years—including, in particular, the links between coal-fired power generation, air pollution, and public health. And, of course, the climate-and-economy linkages have become more prominent, as emphasized in the work of the Global Commission on the Economy and Climate.

Taken together, this constitutes very substantial change in the eight years since *The Stern Review* was published. In summary, the arguments that the costs of inaction greatly exceed the costs of action, strong then, are still stronger now, and we have deepened our understanding of the dynamics of economic change and international interactions. All that said, notwithstanding substantial progress in many countries, progress is far too slow.

This is a time to choose. The first option involves continuing to rely on past technologies, methods, and institutions: it could give us a sort of growth for a while, in a pattern we know, and which many find unattractive and troubled, which will lead to chaos, conflict, and destruction toward the middle and second half of this century. The second option involves embracing and harnessing the positive changes unfolding around us, investing strongly and innovating intensely to bring about not only a much more attractive way of living but also growth which can be sustained. The choices we now face present an enormous opportunity. But delay is dangerous. If we fail to take this opportunity and attempt to follow the old ways, the opportunity will be gone. We use it or lose it.

That we have such a choice follows from the basic four arguments that I have set out in this book. First, we are on a path with strongly rising greenhouse gas emissions which could lead to global average surface temperatures not seen on this planet for millions or tens of millions of years. The consequences could include hundreds of millions of people having to move, with the associated risks of severe and extended conflict. Second, to avoid these risks, or to reduce them radically, fundamental change will be necessary, including essentially zero emissions in the second half of this century, zero-carbon electricity by midcentury, and managing our forests and land much better than we have in the past. Much of the fundamental change required in emissions will occur in a period of remarkable structural transformation, full of opportunities for efficiency and emissions reductions. That transformation will happen in some form or other: it is we who will determine whether it goes well or badly. Third, delay is dangerous because flows of emissions build stocks of greenhouse gases which are, particularly for carbon dioxide, long-lasting and difficult to remove at scale, and because high-carbon capital and infrastructure, which can last for decades, can lock us into high-emission activities. Fourth, the alternative paths, the transitions to a low-carbon economy, are likely to be full of discovery, innovation, investment, and growth, much as we have seen in other waves of technological change over the last 250 years. Further, the alternative ways of producing, consuming, and living will likely be cleaner, quieter, safer, more energy-secure, more community-based, and more biodiverse in the short and medium term as well as involving far lower climate risks in the medium and long term.

Establishing these arguments on the basis of principles and evidence from science, economics, economic history, ethics, and other disciplines has been the primary purpose of this book.

Getting the arguments right is a crucial necessary condition for motivating the climate action that these arguments justify. However, it is not sufficient: the reasons for inaction go beyond simply the proliferation of bad arguments and the misunderstanding of good ones. Though I and many others find that those arguments provide a compelling case for strong and urgent action, we are, as a world, moving far too slowly. If action is to be accelerated and the grave risks are to be radically reduced, we must understand why we are moving so slowly. In this concluding chapter, drawing on analyses and arguments in earlier chapters but also introducing some new ones on how ideas change, we examine some of the key reasons for inaction and suggest some ways in which they could be overcome. I first set out what I think some of the key challenges are and then discuss some historical examples of how ideas have changed on other relevant issues, and draw lessons that could usefully be applied to climate change. The chapter concludes with some thoughts about whence the necessary societal change could come.

Why Are We Moving Too Slowly?

Analytical Difficulties and Failings

This book has tried to tackle directly some of the analytical controversies—both misunderstandings and deliberate falsehoods—that contribute to the slow pace of action on climate change, including failure to grasp the four key arguments summarized above. Other crucial analytical problems include the unwillingness of many to recognize moral responsibilities toward future generations, the importance of equity among people (noting that it is usually the poorest, who have contributed the least to creating the problem, who are hit the earliest and the hardest), the nature and scale of what is happening around the world, and the important role that international cooperation, of the right kind, can and should play in the response to climate change.

I have also examined some of the analytical weaknesses of arguments made by those who have attacked the science. A common strategy of deniers and skeptics is to try to sow doubt in the mind of the public about the underlying science, which identifies the great risks we face, in order to suggest that climate action is unnecessary or unwise. But even if there were grounds for doubt about the basic science, to draw from such doubt the conclusion that we should not act would require the assumption that we can be confident that the risks are small. It is incumbent on the deniers, particularly because the science points to the dangers of delay, to substantiate that assumption by demonstrating that we can indeed be confident that the risks are small. In the face of the scientific evidence, such a demonstration would be close to impossible. Yet most of the deniers have either failed to understand that the position they espouse requires this demonstration or they are not engaging in public argument in good faith.

Communication Deficit

The logical arguments presented above, if they are to have traction, require effective communication. Action on climate change has been hampered by a deficit in the communication of sound arguments, and a surplus of effective communication of the misguided ones.

Effective communication on climate change requires at least three things.[1] First, the key elements of the analytical case for climate action must be presented together: inspiring change requires an articulation of the problem, demonstration that there are effective and attractive responses, and a sound path for implementation. However, many advocates have focused narrowly on the science and risks of climate change; until relatively recently there has been insufficient communication of the opportunities and benefits of action and the means by which they can be realized. Second, messengers matter. Different audiences trust different types of messengers. If movement for change is to be widespread and gather momentum, we can expect to see different communicators and champions for different audiences. Third, climate change communication, to be effective, must utilize rhetoric and frames that resonate with the values and emotions that could inspire action—in this case, local and global collective action on a large scale.[2]

One or more of these elements is often missing in communication by those who argue for strong climate action. Meanwhile, deniers and other opponents of action have often communicated the arguments for inaction much more effectively, frequently through appeals to self-interested values and mobilizing messengers who are trusted, it seems irrespective of the quality of the case being made.

The media inevitably play a vital role in communicating the risks of climate change and the opportunities for and benefits of action. A study of factors influencing concern about climate change in the US over the period 2002–2010 concluded that "media coverage of climate change directly affects the level of public concern. The greater the quantity of media coverage of climate change, the greater the level of public concern."[3]

Media coverage can be particularly important where issues of frequency, risk, and probability are involved. This is because people tend to assess the frequency or probability of an event by "the ease with which instances come to mind," and personal experiences, dramatic events, and more frequent or prominently reported or observed events come more easily to mind.[4] Since many changes wrought by climate change occur gradually and imperceptibly, and since most people most of the time do not personally experience the extreme weather events that are consistent with climate change, the extent to which an issue is represented in the media can have a large effect on people's perception of its frequency or probability. Such extreme weather events can, indeed, serve as powerful opportunities to illustrate the risks associated with climate change. Yet too often media discussion of extreme weather becomes bogged down in the question of whether it can be proved that the particular event was caused by climate change.[5] The important issue, however, is that climate change is a key determinant of increased risks of extreme events. When extreme events occur, the media have a particularly important role to play in communicating the science accurately, in terms of changing weather *patterns* and *risks* and thereby helping the public to understand the origins of the risks.[6]

The media can also influence our response to events when they do occur. For example, media reportage can affect the extent to which people provide aid to victims in times of crisis.[7] Responses to the Indian Ocean tsunami of December 2004 and super typhoon Haiyan in 2013 appear to have been greatly influenced by seeing the devastation on television screens across the world. One study found that the climate change imagery used in newspapers affects people's perceptions of the issue's importance and of their ability to do something about it.[8]

The importance of frequent, accurate, clear, and accessible public discussion of climate change places a great responsibility on media organizations. However, many such organizations have operated under a misguided conception of what is required by "balance," i.e., that scientific evidence on climate change needs to be balanced with nonscientific opinion.[9] This is clearly a gross distortion, and should be seen as akin to putting flat-Earthers on an equal footing with professional physicists. Even worse, some media have actively pursued an editorial agenda of denial and obfuscation.[10] And focusing on false or misguided debates about the science has arguably diverted media discussion away from discussing possible responses to climate change.

Psychological Barriers

The discipline of psychology has much to teach us about why we are not acting, individually and collectively, with an urgency and at a scale commensurate with the challenge. Increasingly, psychologists are doing applied work focused explicitly on responding to climate change.[11] Psychology is not a discipline in which I specialize, but I have tried to listen to and learn from psychologists, including Danny Kahneman and Bob Cialdini, to whom I am very grateful.[12] I cannot hope to do justice here to the myriad psychological processes that work against action on climate change, but it is worth highlighting a few key lessons.

The discussion on communication highlighted some of the cognitive heuristics and biases that can lead us astray in the perception of frequency and probability. We also know that people's attitudes and behavior can be strongly influenced by situational influences. Surveyed attitudes toward climate change can be affected, for example, by the local temperature at the time and place at which the response is elicited,[13] by whether the respondent perceives the current temperature to be warmer or cooler than usual,[14] and even by the temperature of the room in which they happen to be.[15]

There is also evidence that social and cultural features of one's situation have a systematic influence on perceptions of risk generally, and on beliefs about climate change in particular.[16] Dan Kahan and colleagues at Yale's Cultural Cognition Project found that people's risk perceptions concerning climate change correlate with their basic values: low risk perception correlates with an individualistic/hierarchical worldview; high risk perception is more likely if an individual's world view is more communitarian/egalitarian. Additionally, they found that greater scientific and numerical ability has a slight polarizing effect: those with an individualistic/hierarchical worldview perceived climate change to be of *less* concern the greater their scientific and numerical ability. Interpreting their results, the authors posit that people adapt their beliefs about climate change to conform to their social peer group; and that people are more proficient at so adapting the more scientifically literate and numerate they are.[17] If this interpretation is correct, then it would suggest that the task of communicating and persuading people to act on climate change may be more challenging than many believe. Different "messengers" will have different effects for different audiences.[18]

Another barrier to climate action—and indeed to any kind of structural change—relates to the disproportionate value people ascribe to avoiding losses as opposed to achieving gains (people are "loss averse"), and the power of the status quo as a psychological reference point against which people code outcomes as gains or losses (the "endowment effect" and "status quo bias").[19] These phenomena are the psychological roots of a tendency toward social, economic, and political conservatism that makes it more difficult to pursue the alternative, low-carbon path that I have described, notwithstanding the attractiveness of that path relative to the status quo.

Psychological insights are also useful in illuminating why climate actions that are economically rational at a micro scale, such as cost-effective measures to improve energy efficiency, are not taken. One challenge is that the rewards, in terms of energy and cost savings that could be made by simple household efficiency measures, are not immediately clear to people in their daily household activities. Technology and design can help to overcome these barriers through products that make it simple for people to both understand their energy use and adjust it to save money; Google's $3.2 billion acquisition of Nest Labs, a leading product developer in this area, in 2014 indicates their view of the market potential.

Structural Barriers

In addition to barriers at the individual level, we can identify structural and institutional barriers. A first set of structural challenges concerns the organization of politics, and the structure of political economy, within many countries where action is needed most urgently. There will be groups who could see themselves as threatened that can or will act collectively and politically. Industrial revolutions involve dislocation—for example, this one requires the rapid decarbonization of the energy sector, and policy-induced increases in the costs of emissions-intensive goods and services (e.g., hydrocarbon-based electricity), so that the cost includes the damages inflicted on others. Managing such changes, particularly where this involves dislocation, will always be politically challenging—particularly when the costs are short-term, some of the co-benefits are medium-term, and the climate benefits are long-term. Many politicians face short-term electoral incentives and there is a temptation to avoid perceived costs and disruption, which means avoiding action despite the attendant medium-and long-term benefits. We may hope our leaders would act in the longer-term interest, even if this involves political risks. Good leaders can do this well, drawing political energy from pursuing a larger vision. But we must understand that perceived structures of political incentives may point in the wrong direction, and must think carefully about how those incentives can be influenced.

There are deep structural problems in the politics and political economy of many countries that lead to the disproportionate influence of vested interests over the formulation of policy and the tenor of public opinion, along with the mass disengagement of ordinary citizens from the political processes that shape their lives. Some recent analysis suggests that representative democracy in many parts of the world is facing a profound crisis[20] — and at precisely the time it needs to be functioning at its best if it is to take decisions that are important for the long term.

Structural features in the worlds of business and finance, in the media, and in social relations may also hold back climate action. In business and finance, excessively short-term incentive structures direct capital away from long-term value creation,[21] and a dearth of disclosure and transparency requirements in many jurisdictions on key issues like firm-level emissions, energy use, investments in fossil fuel assets, and so on hinder the efficient operation of markets and the democratic process. In many countries the media are not structured in a way that encourages the critical democratic function of promoting long-term public interest. They are often organized on a narrow model according to which many of the key media players see themselves as providing a purely private good to viewers, listeners, or readers, and are oriented to very immediate "consumer interests." That can undermine discussion and information about economic, community, and social interests and the longer term.

There are important difficulties connected with social structures, divisions, and inequalities. For example, it has been found that "conservative white males are significantly more likely than are other Americans to endorse denialist views" on climate change, and that "these differences are even greater for those conservative white males who self-report understanding global warming very well."[22] Moreover, debates in the US about energy and carbon pricing are made more challenging by existing inequalities in income and wealth: those at the top have a disproportionate ability to use their wealth to influence argument and political choice, and are particularly motivated to do so when wealth is associated with hydrocarbons.

Examples from the History of Social Change

The barriers to climate action are many. Generating the necessary change will be fraught with challenges. But societies have made big, difficult structural changes before. How have they done so? There is no very close parallel to a risk-related, policy-based change of economy and society on the scale required by climate change; but there are other major changes that are instructive. I will touch here on the cases of smoking, leaded petrol, drunk driving, and HIV. Later in this chapter I shall examine some relevant examples of social justice movements.

The evidence on the effects of smoking on health was built on the epidemiological work of Richard Doll and others in the 1950s and 1960s, together with an understanding of the biological mechanisms at work. It was developed over a number of decades. Eventually the overall evidence on the risk was overwhelming, and strong public policy came into effect. Regulation, tax, information policies, and advertising were (and still are) all used. Policy was based on evidence. But the process took a long time and was fiercely contested by vested interests. More than 50 years after the first clear results, smoking rates have declined significantly in many advanced countries but are still rising in many low-and middle-income countries, where 80% of the world's smokers now live.[23] Policy on smoking seems to have been led by the medical profession, which, after seeing the evidence grow stronger and stronger, put pressure on governments to act. The profession also shared its concerns with the general public in an attempt to encourage it to change its ways, and also to build support for its arguments to government on the importance of action. The medical profession and government appealed to our "higher selves"—the desires of many smokers to give up smoking—or to our health interests in a more objective sense, over "the lower self" that is more vulnerable to temptation, immediate gratification, and addiction. But evidence on damage to others through "passive smoking" appears to have been important, too.

In many ways, policy and action on smoking followed an expert-led, top-down, professional route to formulating policy and fostering the behavioral change that was required to manage and reduce risk. In that case, strong evidence, effectively communicated, was essential. So too was the willingness to take on the powerful vested interests in tobacco.

The replacement of leaded by unleaded petrol, which has occurred partially or completely in most countries, has been driven by two factors: increasing evidence and concern about the health impacts of lead, resulting in regulatory measures to reduce, and in many countries phase out, the content of lead in gasoline; and the use of catalytic converters, with which leaded fuels are incompatible.[24] These two factors played different roles in different countries.[25] In those countries where health concerns were the driving force, we have seen developments in some respects similar to smoking—an understanding of epidemiology and the biological mechanisms at work led to top-down action, with regulation being the main policy tool. On both smoking and unleaded petrol, once policy and action gained momentum, they spread fairly quickly. On unleaded petrol the battle was won relatively quickly, and leaded petrol is now very rare around the world. On smoking it continues.

Policy change on drink and driving appears to have had stronger bottom-up pressures. In the US, Mothers Against Drunk Driving, or MADD, appeared to have a powerful influence.[26] In the case of tobacco, the dangers of passive smoking showed that smoking was not just a decision for individuals about their own fate, and with drunken driving the evidence of risk to others was still more obvious. Thus for both drunk driving and smoking, highlighting the danger to others may well have played a powerful role in combating a more individual-choice-based or narrowly libertarian position.

Interestingly, on drunk driving the changing behavior came about in at least two ways: the developing understanding of the irresponsibility of drunk driving in its dangers to others, and the incentives or penalties such as fines, driving bans, and imprisonment. The legal imposition of significant penalties seems to have reinforced the emergent social norm and moral stigma associated with drunk driving. Changing perspectives on irresponsibility appeared to come from overall evidence, knowledge of individual cases (either through personal life or in the media), and public discussion of the issues.

Action on HIV/AIDS seems to have been still more strongly driven from below than in the case of drunk driving. Those who demanded action to tackle HIV/AIDS felt that their goals would not be addressed on the scale and with the urgency necessary unless they conducted a vigorous and highly political grassroots campaign. They had to deal, in some parts of the political spectrum, with a moralistic opposition or reluctance to act associated with censorious views around sexually transmitted diseases. The campaign by groups such as ACT UP was high-profile in the US and elsewhere, with targeted demonstrations and campaigns including on Wall Street and at the offices of the Food and Drug Administration and the National Institutes of Health. It was very effective at both national and international levels in mobilizing necessary public health action, treatment, and research. There is, of course, much more to do on HIV/AIDS around the world, but there is surely a strong lesson here about effective public grassroots pressure.

Generating strong action on climate change is a challenge that exhibits features not shared by these other issues. Tackling climate change is not merely about reducing the use of a single type of product, or treating one single source of risk; it will require setting in train a dynamic process—an energy-industrial revolution—that fundamentally transforms our economy and society. Industrial revolutions and waves of technological change come about in different ways from those associated with the pressures arising from public perceptions. They are driven by anticipated returns from investment and alternative activity or occupations. Demand for new products or services (electricity, motor cars, trains, communication, and so on) is influenced by consumer recognition of their usefulness or value. The lesson here is surely that if new technologies and different ways of doing things are to advance quickly, they must be seen as profitable, attractive, and useful. In the case of climate change it will be, in part, policy that will make them so—for example, by influencing the behavior of producers and consumers with prices and taxes that reflect the costs of emissions, and thus making markets work better. That is a key difference from other industrial revolutions.

At the same time, we should not see policy only in terms of price and market incentives. We have seen in the above examples and earlier in the book that regulation can have a very powerful role to play, as it did with unleaded petrol. So too can social and moral norms, which change perceptions of what is morally and socially responsible behavior by individuals, businesses, and governments. As such, the process of change associated with climate action can also learn from past revolutions in moral and social attitudes—for example, the abolition of slavery; the decline of the practice of foot-binding in China; the decline of dueling and other forms of violence; the expansion of voting rights and other civil rights to historically oppressed groups; and, more recently, international divestment from apartheid South Africa.[27]

The Lessons for Climate Change

Let me try to distil some lessons from the examples considered above, and from others touched on or explored elsewhere in the book (e.g., the past waves of technological progress discussed in chapter 2).

Good analysis is critical. In the cases involving policy responses to risk, such as smoking, lead-ed petrol, and HIV/AIDS, good analysis played a critical role. Generating policy change requires, first and foremost, posing the policy questions in the right way—in other words, in a way that reflects the issues we face, and is based on sound theory and evidence. Climate change concerns the management of risk on a colossal scale. Second comes the marshaling of the arguments and evidence about the risks so that there can be an understanding of what is at risk and what is involved in reducing those risks—on climate change, this spans the reduction of emissions and adaptation to the climate change that is now unavoidable. In other words, the identification of options for action that can tackle the issues. Third, we have to examine the details of alternative paths and show their benefits, costs, and necessary investments.

Appealing to values and a sense of justice can be powerfully motivating. In many examples of historical change—the HIV/AIDS case, for example—appeals to values and a sense of (in)justice have been powerful motivators for action. Two simple but crucial points stand out from the arguments presented in this book. First, climate change involves the causation of harm, and risks of harm, to current people and to future people (and to many other living things) on a massive scale. Perspectives from each of the diverse ethical theories considered in chapter 6 would likely regard this harming as a wrong. Framing emissions in terms of causing harm may be more morally motivating than appeals to utilitarian calculations. Personal involvement in the causation of harm engages people's emotions in a way that impersonal calculations do not.[28] This phenomenon may partially explain why, historically, many powerful moral campaigns have involved demands to stop causing harm.[29]

Second, the transition must be equitable and seen as equitable. That requires the global economy to be decarbonized in a way that promotes development, growth, and poverty reduc-tion in poor countries. Decarbonizing the global economy equitably is eminently possible and is the moral core of equitable access to sustainable development, in the language of COP 16 (described in chapter 9). Similar principles apply to considerations of inequality and poverty within countries.

Communicate strategically and use examples. Effective communication was very important in the case studies of change discussed above, particularly those relating to smoking, drunk driving, and HIV/AIDs. In my experience of climate change communication, examples are often very powerful devices, particularly in demonstrating the impacts and risks and in showing the attractiveness of a low-carbon path. In developing countries in particular, if climate action is to be accelerated, the argument must be convincing that an environmentally sustainable growth path is not a threat to the fight against poverty. There are communities and countries that are showing the way, as we have seen in this book, but examples have to be shared and multiplied. Thankfully, eight years on from *The Stern Review*, there are many more examples that can be mobilized to communicate the benefits of climate action.

Extreme weather events can be the most powerful examples of all. They can provide "moments of power"[30] to make the case for climate action, as then Mayor Michael Bloomberg of New York, and others, found with Superstorm Sandy in 2012. A prime minister of Australia without the prejudices against climate science of Tony Abbott would have seen the extreme heat and fires of 2013–2014 as an opportunity to do the same. Let me be clear that this is not about faulty arguments such as attributing with certainty a particular extreme weather event to climate change. It is about using examples to illustrate patterns and dangers and showing that such events could become more severe and more common—and, crucially, explaining that what we see now, at about 0.8°C above the nineteenth century, is tiny relative to what we risk at 3, 4, or 5°C.

Governments might also appeal to the appropriateness of people paying the full cost of their actions when they buy goods, emphasizing that carbon pricing and regulation are simply a means of ensuring better-functioning markets. People might recognize the sound policy in abolishing the subsidy that is associated with, or defined by, the ability to emit and pollute for free. Governments might also appeal more directly to people's sense of responsibility for avoiding harm. There is some analogy to the desire to avoid goods made by child labor, or to buy fair-trade tea even though it might cost a little more. And the linkages between international climate and development policies can be strengthened, rhetorically as well as substantively, so that mitigating climate change is, and is seen to be, a core feature of development, intimately intertwined with poverty reduction. Indeed, climate responsibility is about the sustainability of poverty reduction in the future, and the action involves fostering inclusive growth and poverty reduction now.

Packaging policies. From the case studies of previous social change, we can see that a wide variety of creative interventions by governments can have an important direct and indirect effect in achieving changes. This book has set out the combination of the most important policies for tackling climate change. However, given the many constraints, both structural and transient, that governments face in trying to undertake serious reforms, these policies will need to be supplemented, supported, and packaged in strategic and creative ways, utilizing all means that governments have at their disposal, and tailored to local conditions.

In the smoking, HIV/AIDS, and drunk driving examples, governments in many parts of the world played helpful roles by providing information, nudging people by making it easier to undertake more healthy behavior, using their rhetorical powers to persuade the public, and in many countries effectively using emotionally resonant public advertising to shift attitudes and preferences. Many of these levers could usefully be deployed in tackling climate change alongside the main, more economics-oriented policies discussed in this book.

There are other ways that politicians can package and frame climate action to overcome the difficulties they face. Some climate action does have short-term benefits: for example, "green stimulus" in recessions, when interest rates are low and labor is underutilized, can boost growth, jobs, and incomes; and energy efficiency measures can reduce energy bills immediately. Moreover, stimulus of a kind that promotes the growth story of the future is surely wiser than one that tries to promote activity in areas that are damaging and will decline. Reduced air pollution from phasing out coal, for example, which has very large potential benefits, could come through quite quickly—recall that the social costs of air pollution are more than 4% of GDP in many countries, and more than 10% in China (see chapter 2).

Integrating climate action with other reforms, such as lowering other taxes, increasing health services (for example), other efficiency-improving reforms, sound industry policy to support the growth of low-carbon industries, investments in public transport, and so on, with clear articulation of the climate benefits and co-benefits, can also help to mobilize support for climate action.[31] If the reforms are efficient, by definition the gains to the winners of reform should outweigh the losses to losers and it should be possible to identify and cultivate a coalition of support for the changes. Governments can also play a powerful convening role in fostering such coalitions to support reform efforts. The benefits of this type of reform, focused as it is on efficiency, growth, and risk reduction, should in principle be supported by a financial and business community that is thoughtful and analytical.

Carbon-related fiscal measures and accompanying transfer payments as appropriate, together with direct public policies, can help reduce poverty and inequality. For example, ensuring that some of the revenues from carbon/environmental taxation are applied to assisting affected workers and communities find new sources of employment and income in the low-carbon economy is both a moral and a political imperative. And part of the revenues could be used to protect poorer people from any associated rise in energy prices. Giving people time to adjust to changes in prices by phasing in carbon prices, or abolishing fuel subsidies, gradually and in ways that can be predicted is another way of managing distributional impacts. Taking the opportunity to phase out fuel subsidies during periods when the world price of oil is falling makes political sense. One of the most effective ways of tackling fuel poverty is through the insulation of homes of poor people.[32] Further, if the dynamics of learning go well, then rises in energy prices as a result of climate policies are likely to be temporary.

International cooperation can help drive change. International cooperation has helped drive political change across a range of historical issues, and it will be very important in the case of climate change, as we saw in part III of this book. First, it is important to understand what others are doing and planning. That is the foundation of cooperation. Without that understanding, it is all too easy to assume that they are doing nothing or very little. Second, the plans a country indicates should be credible if they are to form the basis of cooperation. That does not necessarily mean that they need to be "legally binding and enforceable" at the international level; rather, the credibility of a country's plans is primarily a function of its domestic structures, institutions, understandings, and track record. Third, international cooperation and domestic actions can reinforce each other. Confidence in the latter leads to progress in the former and vice versa. It is a great mistake to assume that we have to give up on the former and rely only on the latter. And fourth, we must think rigorously about the *type* of institutions and principles at the international level that are best suited to building confidence and driving domestic structural transformations.

The pace of transformation to a low-carbon world is unlikely to be steady: build momentum. Big changes take a long time to initiate, but once initiated can take hold surprisingly quickly. Decarbonizing the global economy will not be a simple process that involves applying pressure and then straightforwardly peaking and reducing emissions at steady rates until they reach zero or near zero. As the last quarter-century of efforts to tackle climate change has shown, political, social, technological, and economic change of this magnitude can be a long time coming; much effort can be spent with seemingly few results. This can be dispiriting. But big changes can happen very quickly once tipping points are reached, whether social, political, economic, or technological. A wave of low-carbon innovation, growth, and prosperity will likely develop a momentum of its own, and the low-carbon world will become the normal one. The challenge is to make this happen sooner rather than later.

Technological, economic, social, and political change are all needed and can reinforce one another. Pressures and forces from a variety of directions are typically needed to generate large-scale changes. The interaction between developments in science/knowledge, professional engagement, policy leadership, and grassroots activism were all important, to various degrees, in the cases of smoking, drunk driving, leaded petrol, and HIV/AIDS discussed above. With regard to technological revolutions, technological and economic conditions are typically important drivers, as we have seen. However, changes in social, institutional, and political arrangements can change perceptions of what is possible and desirable, create new markets, generate demand for new products, services, and business models, give rise to new types of skills and knowledge, allocate capital, and otherwise change the incentive structures within which technological and economic forces operate.

On climate change, we already see technological innovation, changes in relative prices, changes in social norms, new policy interventions, and other pressures having a significant effect. These forces are not isolated; they interact. German feed-in tariffs, and Chinese economic conditions, led to the large growth in installed capacity of solar PV, which brought down costs through learning and scale, changing relative prices in many countries, leading more people to put solar panels on their roofs, spurring the growth of new industries, changing the political economy, creating new pressures on and opportunities in electricity markets, affecting business models, generating pressures for further policy change, and so on.

Leadership and Social Pressure: Likely Sources of Change

The acceleration of action on managing climate risks requires policy change, and we must therefore examine how such change can happen. Leadership and social pressure are critical.

Leaders who are respected and trusted, and can communicate the issues clearly and effectively, have played important roles in social and policy change throughout history. Leaders have often catalyzed social movements for change, and, conversely, such movements have generated new leaders. What might the sources of leadership and social pressure be?

Those in political leadership carry a special responsibility for the future of their country and thus a responsibility to take a long view. Unfortunately, trust in politicians is low in many countries, and many politicians who understand the issues have been diverted by economic crises or intimidated by confrontation with vested interests. Strong action on climate change is often seen as making short-term election more difficult. But there is no more important issue, and it is their duty to lead.

Pressure from civil society, the business community, and subnational governments can foster national political leadership. In many societies, religious leaders and movements will be prominent in influencing social and political attitudes and decisions. In my view, they are of particular significance on climate, since at the heart of the arguments are ethical issues around moral responsibility toward younger generations and moral responsibility toward the world as a whole. And some of the most prominent religious leaders are beginning to raise the issue: Pope Francis has made the case for climate change action and environmental protection; when speaking on these issues he warned an audience in Rome on 22 May 2014 that "if we destroy creation, creation will destroy us!"[33]

In many countries royal families carry special respect, and that too places a responsibility for balanced and considered leadership. Royal families often take a long view—many see their role as, in large measure, the promotion of the long-term welfare of their country and its peoples. And the lifetime of a dynasty is much longer than that of a government. There are some strong examples, such as H.R.H. Prince Charles in the UK, of royals who accept that responsibility and speak out strongly and effectively.

Some viewers, listeners, and readers have strong trust in particular media outlets—for example (to take a UK-focused view), the BBC in the UK and the BBC World Service worldwide. Such outlets carry a special responsibility to present the issues regarding climate change in a responsible way. There are prominent individual broadcasters or writers who carry great trust, such as Sir David Attenborough or Lord Melvyn Bragg, to take UK examples again. They too carry great responsibility; in these two cases they bear it well.

Actors, celebrities, and sports stars are followed by many, and they also have their responsibilities. It is interesting that these are the people whom UN agencies often enlist to be "ambassadors" for important causes. In many cases, young people look up to them. And young people can mobilize them. The example of Kony 2012 showed how empathy by young people across the world toward the young forced into Joseph Kony's "Lord's Resistance Army" in East Africa could generate pressure for celebrities to demand action. Another example has been the role of Bono and Bob Geldof, through major concerts such as Live Aid in the 1980s and Make Poverty History in 2005, in generating support for poverty reduction and other initiatives in Africa.[34] The latter generated strong pressure on G8 leaders at the July 2005 Gleneagles summit, and that meeting produced a substantial debt relief package.

Academics, teachers, and professionals also have their roles to play: they are often (though not always) seen as trustworthy. We are familiar with the role played by scientists: the US National Academy of Sciences and the UK's Royal Society have been outspoken, as have scientific academics across the world. There is much more to do in communicating, but their focus is increasing and they are becoming more effective.

The medical community, in particular, enjoys high levels of trust and respect in societies across the world. There are many medical professionals involved in public health who are calling for action on climate change. We have seen in the last few years the mounting evidence of the dangers of air pollution from the burning of fossil fuels. Already this accounts for millions of deaths a year. Recent calculations of costs[35] in terms of GDP based on WHO health evidence point to costs in China of more than 10% of GDP and in Germany of 6%, with high costs in many other countries.[36] It is a major issue. Other important potential health costs include extreme weather events, malaria, communicable diseases, and the movement of people, particularly from potentially severe and extended conflict. And alternative lifestyles such as those involving less urban pollution, more walking and cycling, and better public transport can bring not only higher material standards of living but also better health. Encouragingly, there is a growing literature, and increasing political and policy mobilization, among the health community for action on climate change. I mentioned in chapter 7 the Climate Health Commission, which extends pioneering work done in 2009 by the Lancet-UCL Commission on Climate Change.[37] The health dangers of climate change also received prominent attention thanks to interventions by the health community around the time of the latest IPCC reports.[38]

In addition to these sources of pressure, there are three further sources that are likely to be especially important: business, cities, and young people.

Among the most important of the constituencies are businesses, from small farms to big international corporations. Notwithstanding problems of short-termism in some parts of the business world, many businesses remain inclined to take a long view, and those businesses can be very influential. We have seen throughout this book how businesses can lead through the power of their example, including by producing low-carbon and otherwise sustainable goods and services, by making big advances in their energy and resource efficiency (see chapter 2), by promoting emissions reductions and environmental sustainability throughout their supply chains, and by applying an internal carbon price to their operations (see chapter 7). In these ways and more, businesses are demonstrating what can be done and how to combine growth and environmental responsibilities.

Business can also play a powerful and constructive role through the advocacy of strong and clear climate policy. While it is all too often the incumbent beneficiaries of the high-carbon status quo who dominate politics and policy formation, public business leadership on climate change is gathering pace. Caring for Climate—a joint initiative of the UN Global Compact, UNFCCC, and UNEP—seeks to mobilize a critical mass of companies around the world to demonstrate leadership on climate action. In accordance with the initiative's Business Leadership Criteria on Carbon Pricing, businesses are invited to "publicly advocate the importance of carbon pricing."[39] More than 1,000 businesses and investors—alongside 73 national governments, 11 state and provincial governments, and 11 cities—signaled their support for carbon pricing through a series of initiatives announced at the UN Secretary-General's Climate Leadership Summit in September 2014.[40] In Europe, the Prince of Wales's Corporate Leaders Group called for governments to put in place policies to prevent the cumulative emission of more than a trillion tonnes of CO_2, arguing that passing that threshold would lead to unacceptable levels of climate-related risk.[41] And We Mean Business, a coalition of organizations working with thousands of the world's most influential businesses and investors, advocates that governments implement a range of climate policies, from eliminating high-carbon subsidies to carbon pricing, and adopt a long-term global goal of "net zero emissions well before the end of the century."[42]

These examples show that many businesses understand the risks of climate change and the impacts it will have on their businesses; they see the market growth opportunities of a low-carbon world; they recognize that customers, shareholders, and staff look for environmental responsibility in their activities; and they are seeking clear signals from governments so that they can invest with confidence in the low-carbon economy.

Cities, too, can be powerful agents of change. First, cities can see many of the co-benefits of climate action in a very direct way. Better city planning, public transport, walking and cycling infrastructure, and urban green spaces reduce congestion and urban air pollution, improve mobility and the efficiency of travel, and create a more appealing urban environment. Green buildings save energy, water, waste, and materials, and hence costs and resources in general, and bring home to people the benefits of green design in a very powerful way. Second, with some 90% of urban areas situated on coastlines, cities are on the front line of climate change, vulnerable to impacts such as sea level rise and coastal storms[43]—and their citizens share the impacts of disasters together as a community in a way that can be a powerful motivator for change. Consider the examples of New Orleans after Hurricane Katrina in 2005, the Mumbai floods of the same year, and New York after Superstorm Sandy in 2012. Third, cities are big enough to be very significant in terms of both emissions and politics. Cities house more than half the world's population, consume over two-thirds of the world's energy, and produce 70% of global CO_2 emissions.[44] And, while city governments' jurisdictional authority varies across the world, cities commonly have significant powers over transport, buildings, waste, energy efficiency, urban finance and economic development, community development, and adaptation.[45] And fourth, city governments have an advantage in policy implementation because they are typically physically closer to, and have deeper relationships with, their constituent residents and businesses than do officials and agencies at the state or national level.

Cities like Singapore and New York provide strong examples of what cities can do, and networks like C40 Cities are helping to diffuse good practices and build collaborative initiatives across large cities (see chapter 7). Through a combination of leadership and social pressure, city governments and their constituents could well push their national counterparts toward more ambitious climate policy.

Finally, young people are, and will continue to be, a powerful source of pressure for climate action. For it is they who will suffer most from the negligence of earlier generations, including this one. When I was young, it was apartheid, Vietnam, and civil rights that moved many of us to protest and agitate for change. And change came. There are differences, but also numerous parallels, with the issue of climate change. In earlier cases, the changes being demanded could be produced (once the political will had crystallized) more quickly and decisively than the changes needed to tackle climate change. On the other hand, those cases required long and difficult struggles led by those most deeply affected but where others could act in support. The common object of those struggles was to overcome destructiveness and injustice. Climate change is destructive and unjust.

Today's young people can and should hold their parents' generation to account for their present actions. They can elicit an emotional response that can motivate action. If thinking about the lives of unborn future generations seems too abstract to motivate you to act, try instead looking a young child or grandchild in the eye and asking yourself what sort of future you are leaving for them. There is something that, on reflection, many adults would surely find repugnant in the idea that they will leave their children a damaged planet that will radically affect their life possibilities.[46]

Of course, while many of the ideas and much of the science of climate change are long-standing, it would not have been part of the schooling of most adults. Schools have a special role to play in building a measured, evidence-based understanding of climate change, and in fostering discussion and reflection about the sort of values that their societies should uphold and pursue. Children can then teach their parents: I am reminded of the song "Teach Your Children Well" by Crosby, Stills, Nash & Young,[47] which also says "teach your parents well." Education goes both ways.

This book has sought to show that the case for urgent and radical action is extremely strong, and that the tools to make it happen are firmly within our grasp. To generate the acceleration that is now critical if we are to avoid dangerous climate change, we need leadership and social pressure to keep building—from young people, from cities, from business, from all of the other sources I have described, and more. If that pressure is focused, intelligently and vigorously, in the right areas, if governments heed the lessons, and if well-designed policies are implemented as a result, then political tipping points could be reached and big changes could happen surprisingly quickly.

We are at a remarkable point in history. We have a chance to combine the profound structural changes we are seeing in the world economy and extraordinary technological change on the one hand with a rapid transition to a low-carbon economy on the other. We can simultaneously find a much more attractive way to grow and develop, overcome poverty, and radically reduce the grave risks of climate change. We must decide and act or the opportunity will be lost. The time is now.

Why are we waiting?

Notes

1 The science of psychology has yielded many insights into the psychological mechanisms at work in the course of communication, some of which I discuss further below. Yet our understanding of the basic elements of effective communication, at least in spoken form, can still learn much from Aristotle's treatise *Rhetoric*, from the fourth century BC: it requires (i) ethos—credible, trusted messengers who can combine (ii) logos—appeals to logic with (iii) pathos—appeals to emotion.

2 See, e.g., Kasser and Crompton (2009); Corner and Roberts (2014); Marshall (2014).

3 Brulle, Carmichael, and Jenkins (2012). The study also found that two other factors were important influences on Americans' concerns about climate change: the prominence of other issues (economic downturn, foreign wars); and "elite cues" (i.e., from political leaders), with the latter being the most important factor.

4 Kahneman (2011) calls this the "availability" heuristic. In a famous study conducted by Paul Slovic and colleagues (Lichtenstein et al. 1978), respondents were asked to compare the frequency of various pairings of "causes of death"—one frequently reported in the media, the other not. Respondents' answers were systematically, and incorrectly, biased in favor of the media-reported causes—for example, tornadoes were seen as more frequent killers than asthma, though the latter caused 20 times more deaths. See discussion in Kahneman (2011) at p. 138.

5 Although it should be recognized that direct causation can sometimes be established.

6 Painter (2013).

7 See, e.g., Oosterhof, Heuvelman, and Peters (2009).

8 O'Neill et al. (2013).

9 Boykoff and Boykoff (2004).

10 Interestingly, climate change skepticism in the media appears to be a largely Anglophone phenomenon: Painter (2011).

11 For a good survey of the relevant issues and literature, see American Psychological Association (2010).

12 See Kahneman (2011); Cialdini (1993).

13 Joireman et al. (2010).

14 Li, Johnson, and Zaval (2011).

15 Risen and Critcher (2011).

16 See, e.g., Leiserowitz (2007).

17 Kahan et al. (2012).

18 Ibid.

19 See Kahneman and Tversky (1979); Kahneman, Knetsch, and Thaler (1991); Kahneman (2011).

20 See, e.g., Coggan (2013); Wilks-Heeg, Blick, and Crone (2012); Gilens and Page (2014).

21 See, e.g., Kay (2012).

22 McCright and Dunlap (2011a).

23 World Health Organization (2014).

24 Lovei (1998).

25 Ibid.

26 See http://www.madd.org/about-us/history/.

27 See, e.g., Appiah (2011); Bicchieri (2006); Brennan et al. (2013); Pinker (2011).

28 Greene et al. (2001).

29 Gauri (2012).

30 Cialdini (1993).

31 This discussion of "packaging" reforms draws on Ahmad and Stern (1991) and Green and Stern (2014).

32 Hills (2012).

33 Vatican Radio (2014).

34 I had the privilege of working closely with Bob Geldof when leading the writing of the Report of the Commission for Africa in 2004–2005.

35 Such calculations of loss of life in terms of GDP are standard but can be problematic. The point here is that the loss of life and health costs are very large.

36 See Hamilton (2014).

37 See https://climatehealthcommission.wordpress.com/ucl-lancet-commission/.

38 See, e.g., the front cover and editorial in the 5 April edition of the *British Medical Journal* and the letter to *The Times* by 60 medical professionals on 29 March 2014.

39 See www.caringforclimate.org. The other two actions that businesses are invited to undertake under the Criteria are setting an internal carbon price and communicating progress.

40 See World Bank (2014c).

41 Business Green (2014). See section 1.4 of this book for a discussion of carbon budgets and the remaining carbon space.

42 See http://www.wemeanbusinesscoalition.org/.

43 C40, http://www.c40.org/why_cities.

44 C40, http://www.c40.org/why_cities.

45 C40/Arup (2014b).

46 There is a powerful moral argument here. Further, some psychologists argue there is strong evidence that the idea of reciprocity—others helped us so we should help others (Cialdini 1993)—is often persuasive in shaping action.

47 This song, written by Graham Nash and first released on the Crosby, Stills, Nash & Young album *Déjà Vu* in 1970, has a particular resonance for me, as it was played at the ceremony at which I received the 2013 Stephen H. Schneider Award for Outstanding Climate Science Communication. When it came to "teaching our children well"—along with teaching a great many adults—about the science of climate change, there was no one better than Steve Schneider.

References

Ahmad, E., and N. Stern. 1991. *The Theory and Practice of Tax Reform in Developing Countries*. Cambridge: Cambridge University Press.

American Psychological Association. 2010. "Psychology and Global Climate Change: Addressing a Multifaceted Phenomenon and Set of Challenges." Report of the American Psychological Association Task Force on the Interface between Psychology and Global Climate Change.

Appiah, K. A. 2011. *The Honor Code: How Moral Revolutions Happen*. New York: Norton.

Bicchieri, C. 2006. *The Grammar of Society: The Nature and Dynamics of Social Norms*. Cambridge: Cambridge University Press.

Boykoff, M. T., and J. M. Boykoff. 2004. "Balance as Bias: Global Warming and the US Prestige Press." *Global Environmental Change* 14: 125–136.

Brennan, G., L. Eriksson, R. E. Goodin, and N. Southwood. 2013. *Explaining Norms*. Oxford: Oxford University Press.

Brulle, R. J., J. Carmichael, and J. C. Jenkins. 2012. "Shifting Public Opinion on Climate Change: An Empirical Assessment of Factors Influencing Concern over Climate Change in the U.S., 2002–2010." *Climatic Change* 114: 169–188.

Business Green. 2014. "BT, Shell and Corporates Call for Trillion Tonne Carbon Cap." *Guardian*, 8 April. http://www.theguardian.com/environment/2014/apr/08/bt-shell-corporates-trillion-tonnes-carbon.

C40/Arup. 2014b. "Climate Action in Megacities 2.0: C40 Cities Baseline and Opportunities Volume 2.0." http://issuu.com/c40cities/docs/c40_climate_action_in_megacities/1?e=10643095/6541335.

Cialdini, R. B. 1993. *Influence: The Psychology of Persuasion*. New York: Quill/W. Morrow.

Coggan, P. 2013. *The Last Vote: The Threats to Western Democracy*. London: Allen Lane.

Corner, A., and O. Roberts. 2014. "How Narrative Workshops Informed a National Climate Change Campaign." Climate Outreach and Information Network, Oxford.

Gauri, V. 2012. "MDGs That Nudge: The Millennium Development Goals, Popular Mobilization, and the Post-2015 Development Framework." World Bank Policy Research Working Paper No. 6282.

Gilens, M., and B. Page. 2014. "Testing Theories of American Politics: Elites, Interest Groups, and Average Citizens." *Perspectives on Politics* (forthcoming, Fall 2014).

Green, F., and N. Stern. 2014. "An Innovative and Sustainable Growth Path for China: A Critical Decade." Policy paper, Centre for Climate Change Economics and Policy and Grantham Research Institute on Climate Change and the Environment, May. http://www.lse.ac.uk/GranthamInstitute/publications/Policy/docs/Green-and-Stern-policy-paper-May-2014.pdf.

Greene, J. D., R. B. Sommerville, L. E. Nystrom, J. M. Darley, and J. D. Cohen. 2001. "An fMRI Investigation of Emotional Engagement in Moral Judgment." *Science* 293: 2105–2108.

Hamilton, K. 2014. "Calculating PM2.5 Damages for Top Emitters: A Technical Note." New Climate Economy background note. http://newclimateeconomy.net.

Hills, J. 2012. "Getting the Measure of Fuel Poverty: Final Report of the Fuel Poverty Review." CASE Report 72, March.

Joireman, J., H. B. Truelove, and B. Duell. 2010. "Effect of Outdoor Temperature, Heat Primes and Anchoring on Belief in Global Warming." *Journal of Environmental Psychology* 30: 358–367.

Kahan, D. M., E. Peters, M. Wittlin, P. Slovic, L. L. Ouellette, D. Braman, and G. Mandel. 2012. "The Polarizing Impact of Science Literacy and Numeracy on Perceived Climate Change Risks." *Nature Climate Change* 2: 732–735.

Kahneman, D. 2011. *Thinking Fast and Slow*. New York: Farrar, Straus and Giroux.

Kahneman, D., J. L. Knetsch, and R. H. Thaler. 1991. "Anomalies: The Endowment Effect, Loss Aversion, and Status Quo Bias." *Journal of Economic Perspectives* 5: 193–206.

Kahneman, D., and A. Tversky. 1979. "Prospect Theory: An Analysis of Decisions under Risk." *Econometrica* 47: 263–291.

Kasser, T., and T. Crompton. 2009. *Meeting Environmental Challenges: The Role of Human Identity*. Surrey: World Wildlife Fund.

Kay, J. 2012. "The Kay Review of UK Equity Markets and Long-Term Decision Making." https://www.gov.uk/government/uploads/system/uploads/attachment_data/file/253454/bis-12-917-kay-review-of-equity-markets-final-report.pdf.

Leiserowitz, A. 2007. "Communicating the Risks of Global Warming: American Risk Perceptions, Affective Images, and Interpretive Communities." In S. C. Moser and L. Dilling, eds., *Creating a Climate for Change*, 44–63. New York: Cambridge University Press.

Li, Y., E. J. Johnson, and L. Zaval. 2011. "Local Warming: Daily Temperature Change Influences Belief in Global Warming." *Psychological Science* 22: 454–459.

Lichtenstein, S., P. Slovic, B. Fischhoff, M. Layman, and B. Combs. 1978. "Judged Frequency of Lethal Events." *Journal of Experimental Psychology: Human Learning and Memory* 4: 551–578.

Lovei, M. 1998. "Phasing Out Lead from Gasoline: Worldwide Experience and Policy Implications." World Bank Technical Paper No. 397.

Marshall, G. 2014. *Don't Even Think about It: Why Our Brains Are Wired to Ignore Climate Change*. New York: Bloomsbury.

McCright, A. M., and R. E. Dunlap. 2011a. "Cool Dudes: The Denial of Climate Change among Conservative White Males in the United States." *Global Environmental Change* 21: 1163–1172.

O'Neill, S. J., M. Boykoff, S. Niemeyer, and S. A. Day. 2013. "On the Use of Imagery for Climate Change Engagement." *Global Environmental Change* 23: 413–421.

Oosterhof, L., A. Heuvelman, and O. Peters. 2009. "Donation to Disaster Relief Campaigns: Underlying Social Cognitive Factors Exposed." *Evaluation and Program Planning* 32: 148–157.

Painter, J. 2011. *Poles Apart: The International Reporting of Climate Scepticism*. Oxford: Reuters Institute for the Study of Journalism.

Painter, J. 2013. *Climate Change in the Media: Reporting Risk and Uncertainty*. New York: I. B. Tauris.

Pinker, S. 2011. *The Better Angels of Our Nature: A History of Violence and Humanity*. London: Penguin.

Risen, J. L., and C. R. Critcher. 2011. "Visceral Fit: While in a Visceral State, Associated States of the World Seem More Likely." *Journal of Personality and Social Psychology* 100: 777–793.

Vatican Radio. 2014. "Pope at Audience: If We Destroy Creation, It Will Destroy Us." 22 May. http://en.radio-vaticana.va/news/2014/05/22/pope_francis_warns_against_the_destruction_of_creation_/1100782.

Wilks-Heeg, S., A. Blick, and S. Crone. 2012. *How Democratic Is the UK? The 2012 Audit*. Liverpool: Democratic Audit.

World Bank. 2014c. "73 Countries and Over 1,000 Businesses Speak Out in Support of a Price on Carbon." *World Bank Website: News*, 22 September.

World Health Organization. 2014. "Tobacco." Fact Sheet No. 339. http://www.who.int/mediacentre/factsheets/fs339/en/.

Discussion Questions

1   What are the four arguments of Stern's book that he summarizes in this chapter? Are these adequate for orienting you to the problem and direction for action?

2   Many college students are rightfully frustrated by the slow progress of climate action. What are Stern's four reasons for this pace? How do you imagine overcoming at least one of them? Which one(s) related to Oreskes's essay in Chapter 3?

3   Explain whether the examples provided for other social changes have relevance to climate action. Are there other historic changes that you think are more relevant?

4   Explain which of Stern's lessons for climate disruption are most compelling for you.

5   Stern concludes with the importance of different sources of leadership, including youth. Explain which ones you relate to most strongly.

CONCLUSION

In our daily lives it is impossible to see the rapid changes that are occurring with Earth's climate and biodiversity. And in the relatively short time that most of you have been on Earth, it is also hard to detect such changes through personal experience. In the same way we depend on historians to trace how civilization has changed over time—including some periods of rapid change—we depend on scientists to describe what has changed and is changing. The majority of the preceding chapters provide you with an introduction to what scientists know and expect for climate and species.

These readings were chosen to provide you with:

    a.   an introduction to the diversity of life on Earth
    b.   an introduction to the science and effects of climate disruption
    c.   examples of how biodiversity is affected by climate disruption
    d.   a broad perspective on climate disruption, the imperative to address it, and reasons for hope that it will be addressed in time.

You have likely gained an appreciation for the dependency that species have on climate. They evolved with particular climates that had changed more slowly than is presently occurring. The more rapid changes in climate create an additional stressor among many with which species must contend. The evidence of effects on species of changing climate is accumulating and was described in Chapters 7–10. These chapters also described reasonable expectations for the future based on ecological principles and computer modelling of processes that affect species abundances and geographic distributions. The combination of documented declines in some species and reasonable projections clearly make the case for threats to biodiversity due to climate disruption.

The topics discussed in this volume are continuously developing, as they are very dynamic fields of study. Many of the authors continue to research and write on their topics, so you can follow more recent developments through their work as well as the work they cited in their chapters. Every year there are also new researchers and writers presenting new details and syntheses of what is known. I encourage you to continue investigating the topics that interest you most.

Some chapters provide direction for managing landscapes to help species migrate as habitats become less favorable for their existence. As described, these require the integrated action of diverse groups such as academic and government researchers, public land management agencies, non-governmental organizations, businesses, and private landowners. In the final chapter, Nicholas Stern provides broader policy directives and roles of different parts of society for implementing the changes needed to slow the rate of climate disruption. He includes the important role of young people—you—in this process. There is no doubt that the energy, intelligence, creativity, and commitment of your generation is essential to accelerating the actions needed to slow the rate of climate disruption and aggressively address the climate crisis.

CPSIA information can be obtained
at www.ICGtesting.com
Printed in the USA
LVHW100622090120
642813LV00015B/15/P